Gemstones of California

by California State Mining Bureau

with an introduction by Kerby Jackson

This work contains material that was originally published in 1905.

This publication is within the Public Domain.

This edition is reprinted for educational purposes
and in accordance with all applicable Federal Laws.

Introduction Copyright 2015 by Kerby Jackson

Introduction

It has been years since the California State Mining Bureau released their important publication "Gems, Jeweler's Materials and Ornamental Stones of California". First released in 1905, this work has been unavailable to the mining community since those days, with the exception of expensive original collector's copies and poorly produced digital editions.

It has often been said that *"gold is where you find it"*, but even beginning prospectors understand that their chances for finding something of value in the earth or in the streams of the Golden West are dramatically increased by going back to those places where gold and other minerals were once mined by our forerunners. Despite this, much of the contemporary information on local mining history that is currently available is mostly a result of mere local folklore and persistent rumors of major strikes, the details and facts of which, have long been distorted. Long gone are the old timers and with them, the days of first hand knowledge of the mines of the area and how they operated. Also long gone are most of their notes, their assay reports, their mine maps and personal scrapbooks, along with most of the surveys and reports that were performed for them by private and government geologists. Even published books such as this one are often retired to the local landfill or backyard burn pile by the descendents of those old timers and disappear at an alarming rate. Despite the fact that we live in the so-called "Information Age" where information is supposedly only the push of a button on a keyboard away, true insight into mining properties remains illusive and hard to come by, even to those of us who seek out this sort of information as if our lives depend upon it. Without this type of information readily available to the average independent miner, there is little hope that our metal mining industry will ever recover.

Though this volume may not at first seem to be of great importance to gold miners, I feel that those miners with an interest in smelting and refining their finds, especially those recovered from lodes, will find the processes outlined to be of great value.

This important volume and others like it, are being presented in their entirety again, in the hope that the average prospector will no longer stumble through the overgrown hills and the tailing strewn creeks without being well informed enough to have a chance to succeed at his ventures.

Please note that at times it is necessary to rearrange illustration plates in these texts. Any illustrations not found in their original sequence may be found following the index.

Kerby Jackson
Josephine County, Oregon
August 2015

www.goldminingbooks.com

CONTENTS.

LIST OF ILLUSTRATIONS.

GEMS, JEWELERS' MATERIALS, AND ORNAMENTAL STONES OF CALIFORNIA.

By GEORGE F. KUNZ, A.M., Ph.D.

INTRODUCTION.

In preparing this report upon the gem-minerals of California, a few general considerations may be noted at the outset. Prior to the Mexican war, California was a land but little known—a romantic, dreamy region of far-away southern life, where the Catholic fathers had founded their missions, and brought the language and the architecture of Spain to the shores of the Pacific. With the transfer of the territory to the United States, and the discovery of gold in 1848, a swift and mighty change set in, and a wonderful era of progress began, which has continued to the present time.

In this great development of California, there may easily be recognized three distinct steps or stages, the later ones coming in not to replace the earlier, but as successive additions to the productive power of the State. For a number of years California was chiefly known and sought as the land of gold, the El Dorado. Later came the development of the soil for agriculture—the rich harvests of the great valley, and the luxuriant horticulture of the southern section, where the fruits of southern Europe are now gathered on so vast a scale. Lastly, within a few years, has come into view the wonderful richness of the State in gems and precious stones, as a third aspect of productiveness. With this, the present report has to deal.

Many notices of separate observations and discoveries of this kind have from time to time appeared, and lists of minerals found in the State have been published at various periods in different reports of the earlier surveys and by the State Mining Bureau. These will be enumerated further on.* But thus far there has been no general record of the

---

* The titles of many of these reports will be given in the footnotes. See also Bibliography relating to Geology, Palæontology and Mineral Resources of California, including maps, by Capt. A. W. Vogdes, published by the State Mining Bureau, San Francisco, 1904.

distribution of California gem-minerals as such; and the facts have, moreover, been accumulating so rapidly within a very short time past, that they have assumed an importance much greater than before. In 1890 and 1892 the full list of the discoveries known was collected and published with other gem material.*

The discoveries made within the past ten years have been reported almost entirely by the writer, in his capacity as Special Agent of the U.S. Geological Survey for this particular branch, in the annual reports on the Production of Precious Stones in the United States, published in the Division of Mineral Resources, in the annual reports of the Survey. All that has appeared in those volumes, together with a large amount of additional material, earlier and more recent, and much of it from personal communications and other unpublished sources, has been brought together, and corrected up to the date of going to press, so as to present to the people of California a general conception of the wealth of their State in gems and precious stones. It is hoped that this work may give an impetus to further discovery and advance. The developments in this line during the last four years, especially in southern California, are phenomenal, and have not been paralleled by those of any other State or country. They seem, however, only to have commenced, and it is quite possible that in a few years many new and important discoveries will be made in this region, and California be known as one of the greatest gem-producing countries of the world. Hence, while this report is designed to comprise all gem-minerals found in the State, it will deal especially with the newest region of discovery.

It is time, and it is fit, that this body of information should be given to the people of the State; as many of her mineral products are better known and better represented abroad than they are at home. Altogether, probably more than a billion dollars' worth of gold, gems, and minerals of California have been diffused to the ends of the world—the first aiding in commerce and strengthening the banker, the others forming the ornaments and the pride of important museums and other great collections. The tourmaline, spodumene, rock-crystal, and other gems—as familiar now to experts and collectors as the gold itself—have been better known to the residents of Russia, Spain, or Germany, than to the inhabitants of the Golden State whence they came. It is a singular fact that these gems are better represented in the American Museum of Natural History in New York, the United States National Museum at Washington, the British Museum in London, the Musée d'Histoire Naturelle of Paris, and other great institutions in the East and abroad, than they are in the State Mining Bureau of California or the State University at Berkeley—both of which have collections that rank well

---

*Gems and Precious Stones of North America, New York, 1890-92, by George F. Kunz, pp. 31.

for the authenticity and the richness of their specimens, and for their magnificent examples of foreign minerals. If this report shall tend to increase the interest and awaken the pride of the people of the State in their remarkable mineral treasures, it will have served a purpose of permanent utility.

GEM-MINERALS IN CALIFORNIA.

The distribution of gem-minerals in California may be very broadly outlined somewhat as follows:

I. There is, first, the gold region of the central and northern counties along the western base of the Sierra Nevada; in this are found the gold-quartz used so much for jewelry and ornamental work, and the few but interesting diamonds. These latter occur loose in the gold-bearing gravels, sometimes of the surface placers, but generally of the old river-beds now covered and compacted by lava-flows. In these last also is found much of the agatized and opalized wood, which is sometimes capable of use as an ornamental stone. In the same gravel filling of an ancient stream bed, in Calaveras County, has also been found the wonderful deposit of transparent quartz crystals (rock-crystal) of great size, which have yielded some of the finest material for art work ever known anywhere. These occurrences, it is true, are adventitious, and not in the nature of mines that can yield any permanent supply. But they have been found, and may be found again at any time. The gold-quartz is different in this respect, and a fairly steady production of it in certain of the quartz mines may be relied upon hereafter, as before.

The diamonds found in the gravels are neither numerous nor large, but some of them are beautiful and all of them possess much interest. Their occurrence will be described further on in some detail. All have been found incidentally, and no search for them has ever been made. One or two suggestions, however, may be offered here:

(1) As the U. S. Geological Survey is projecting a special study of the occurrence of platinum in California and the Pacific States, it would seem not unlikely that if some attention were paid to the occurrence of the diamond, it also might be found in this exploration; as the diamond is one of the heavier minerals and is almost invariably met with in the riffles with the gold and platinum.

(2) The new grease-board separator used in the South African diamond mines, recently devised through the keen observation of one of its employés—Mr. Kersten of Kimberley—might prove a valuable adjunct to some of the present gold-stamps, or in the sluices, to detect the occurrence of diamonds in California. It is a remarkable fact that while almost all other minerals pass over a board coated with

mutton tallow, when such a board is vibrated or " jigged," all diamonds present remain adhering to the tallow, and can thus be separated. With a contrivance of this kind, diamonds down to the size of a pinpoint are at present saved in the South African diamond washings, while otherwise they would surely be lost.

II.   There is next the region of Tulare County, centering around Visalia, where the recently developed chrysoprase mines occur at several points.   This rare and beautiful stone exists here apparently in some abundance; and associated with it are a number of other forms and varieties of quartz minerals capable of use in the arts for ornamental purposes, such as rose-quartz, chrysopal, etc., besides several species of garnet, some of which have yielded material for gems. Another interesting and rather peculiar stone found in this section, on the borders of Tulare and Fresno counties, is that named by the writer *californite*—a compact green variety of vesuvianite, that perfectly resembles the celebrated ornamental stone known as jade, so much prized in the Orient for elegant art-work.   This is also found in Siskiyou County, at the northern extremity of the State.

III.   The desert region of the southwest, bordering on Nevada and Arizona.   Here, in a country arid, barren, and desolate, consisting largely of volcanic rocks, are found some interesting localities of opal and of turquoise, the latter giving evidence, as in Arizona, of long and extended working by prehistoric tribes, who have left their stone tools and their rock inscriptions around their old places of labor.   These turquoise mines occupy a considerable area in the northwestern angle of San Bernardino County, and are operated by the Himalaya and Toltec mining companies.   The latter company has three groups of mines, all of them patented, situated on the great desert, about 100 miles northwest from Needles station and about 50 miles north of Manvel, which is on a branch of the Santa Fé railroad.   The three mining centers are some 6 miles apart, in the old Solo Mining District, and are known as East Camp, Middle Camp, and West Camp, the latter being within 20 miles of Death Valley.   The altitude is between 5000 and 6000 feet, and as there is no water at either camp, it is necessary to draw it over mountains from 1 to 5 miles.   The same company also operates turquoise mines in Nevada, some 60 miles due east of the others.

The other company, the Himalaya, has a group of five mines in the same district (the Solo), but some distance from the former, being about 60 miles west of Manvel, reached only by team.   These claims are all on one ledge, which is described as a " bird's-eye porphyry" with some granite, striking north and south, with a dip of 75 degrees west.   Turquoise is the only gem found, and occurs in pockets surrounded by a white friable substance, said to be a lime silicate. Two shafts have been sunk to a depth of 80 feet, but no turquoise was found below half that depth; and from there up most of the material has been stoped to the surface. All the work was done by hand, with giant. powder, and was laborious and costly, and has been suspended since February, 1903. During the last year of working, the amount shipped was 431 pounds of matrix and ordinary turquoise, and 49 pounds of picked material.

IV. The most remarkable gem-region of the State, however, is that developed within a few years past in San Diego and Riverside counties, at several localities where lithia minerals occur, among which the gem-tourmalines and gem-spodumenes are especially prominent. Besides these, other gem-minerals have lately been found in adjacent or associated workings, especially topaz, transparent epidote and axinite, pink, green and blue beryl, and essonite garnet—the whole forming an assemblage of such minerals that is scarcely, if at all, equaled anywhere in the world. Many of these mines are as yet only prospects, or trial openings; but the indications are that the region is full of possibilities. Lack of water and fuel are the chief obstacles thus far to a much more extended development.

In general, it may be said that throughout the granitic region of San Diego and Riverside counties there is a widespread prevalence of an igneous rock of gray color which is generally called a diorite, with a little disseminated quartz and mica (biotite); some samples, however, Prof. T. C. Hopkins, whose account is quoted further on, determined to be gabbro rather than diorite. This rock, and the granite, appear in a series of ridges, or mountains, with a prevailing north and south course, and are traversed by dikes, or perhaps, as Professor Hopkins thinks, veins, of pegmatite—very coarsely crystallized granite. These have a general direction of northwest and southeast, and dip southward or southwestward at varying angles at different points. It is in these pegmatite veins or dikes, which vary more or less in their structure, but possess great general similarity, that the gem-minerals are found. In the notes given further on as to the several mines, these special features will be stated in detail.

There are in this region several centers of occurrence, as thus far recognized, of two somewhat distinct types—those yielding lithia minerals, with gem-tourmaline and sometimes gem-spodumene, and those yielding principally garnet, beryl, and topaz. Of the former, three are especially to be noted, in San Diego County—(1) the Mesa Grande mines, which yield crystallized gem-tourmalines of splendid quality, almost exclusively; (2) the Pala district, in which there are three parallel ridges—Pala Mountain on the west, with the great lithia mines and some colored tourmaline; Pala Chief Mountain, in which are found very fine tourmaline and the new and remarkable gem-spodumene (kunzite); and Heriart Mountain on the east, with a number of openings yielding both tourmaline and kunzite; then, northeast of these, in Riverside County, there is (3) the region near Coahuila, in the San Jacinto Mountains; here was the first discovery of gem-tourmaline in California, so far as known to the whites, and kunzite and other lithia minerals have recently been found in association with the tourmaline. There are also other localities between this latter and Mesa Grande, and probably many others may yet be found. About half-way between Mesa Grande and Pala is a fine beryl mine, near Rincon.

The other class of mineral localities appears to lie along a line somewhat southwest of those just noted, extending from near the Mexican boundary, at Jacumba, northwest to Ramona and perhaps beyond, following the general strike of the pegmatite veins, and almost exactly parallel to the line from Mesa Grande to Pala. At Ramona are found abundant fine garnet (essonite), with topaz and beryl, notably the rose variety, but not much tourmaline, no kunzite, and in general little of the lithia minerals. Around Jacumba are found beryl and essonite garnet (often called hyacinth); the latter is abundant, and at one or two points has been worked for several years to some extent. Jacumba, or Jacumba Hot Springs, is close to the Mexican line, some 20 miles east of Campo, and almost on the western edge of the Colorado Desert.

With regard to the character of the rocks, and the relation of the minerals thereto, the following statement from Prof. T. C. Hopkins, of Syracuse University, Syracuse, N. Y., is very instructive. Of particular interest is his distinction between the gem-minerals of the pegmatite veins in the gabbro (diorite) and in the granite. In many of the notes on special features of the several mines, given further on, will be seen phases of the vein-structure that Professor Hopkins here mentions, especially the differences between the upper and lower sides of the vein, which are constantly alluded to. The pockets in which the gem-crystals occur are usually in a somewhat central zone; above is the more typical pegmatite, and some of the describers use that name only for this portion; below lies a finer-grained feldspathic division, which is commonly called the "line-rock," from its being so often lined or banded with minute garnets, as Professor Hopkins describes, or in some cases with minute black or blue tourmalines. The question between gabbro and diorite, for the rock traversed by the pegmatite veins, or dikes, may well be one of locality, as Professor Hopkins concedes in part. The rock may vary in constitution at different points, as it does in physical structure. At one or two places, especially near Dehesa, it takes on the peculiar concretionary character known as orbicular diorite, and this may be valuable as an ornamental stone.

ILL. No. 3. WEATHERED MASS OF "LINE-ROCK," SHOWING GARNET INTERLINEA-
TIONS IN THE COMPACT FELDSPAR. NAYLOR-VANDERBURG MINE,
HERIART MOUNTAIN, SAN DIEGO COUNTY.

ILL. No. 4. ORBICULAR DIORITE, DEHESA, SAN DIEGO COUNTY—WEATHERED SURFACE
SHOWING THE STRUCTURE.

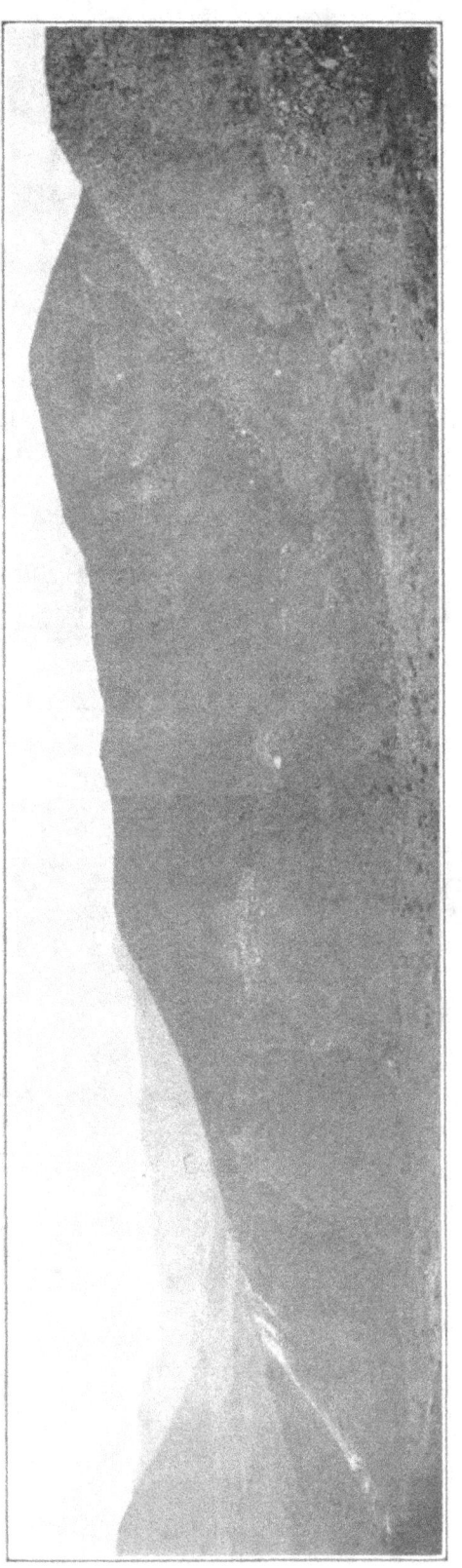

ILL. No. 5. PALA MOUNTAIN—VIEW LOOKING WEST. PEGMATITE LEDGES AS WHITE LINES; DUMP OF THE LITHIA MINE IS SEEN AT EXTREME LEFT.

The Gem Regions of San Diego County, California, by Prof. T. C. Hopkins.— "The gems occur in the midst of pegmatite veins which are formed in the granite and gabbro rocks. These rocks form a range of hills and mountains running through the west and west central portions of San Diego County. Both the rocks occur in great quantities over large areas and meet along very irregular lines of contact, sometimes a long tongue of the granite extending into the gabbro and sometimes the reverse. The gabbro varies considerably in both color and texture. In places it is finely crystalline, grading into basalt with its characteristic columnar structure, and elsewhere it is quite coarse-grained and massive. The granite likewise varies from a coarse-grained texture to a fine-grained felsite and porphyry. In places also it has a gneissoid structure and grades into mica schist. It is a dark gray biotite granite, containing in places dark blotches due to imperfect crystallization of the magma. The pegmatites occur in both the gabbros and the granites, but nearly all of the rich gem-bearing veins are in the first rock. The rich tourmaline and kunzite veins at Pala and Mesa Grande are all in the gabbro. The topaz, garnet, and beryl veins at Ramona are in the granite. The veins are locally very numerous; in some places in the vicinity of Ramona they form nearly half of the rock mass. The gems invariably occur in pockets in the midst of the pegmatite—not always the exact middle of the vein. Generally the vein on one side of the pocket is more coarsely crystalline than on the other side, and in many instances the finer-grained portion is finely banded with small brown garnets. The gems commonly occur embedded in loose clay, but occasionally they are attached to the walls of the pocket and may even be embedded in quartz or orthoclase feldspar. They are almost invariably accompanied by crystals of albite, lepidolite, and quartz. The latter are sometimes quite large, single crystals weighing from 100 to 150 pounds. The country rock has been called a diorite, but a microscopic section from one locality proves it clearly a gabbro, although it may be diorite at other points. The evidence appeared to me pretty conclusive that the pegmatites are veins and not dikes." *

In the summer of 1903, Dr. Waldemar T. Schaller, of the department of geology of the University of California, now of the U. S. Geological Survey, visited the tourmaline and kunzite mine recently opened on Pala Chief Mountain, and reported upon the occurrence in detail. His account of it will be cited further on; at present it is merely mentioned as an authoritative description of this locality, giving the same general facts as indicated over a wider area by Professor Hopkins.†

The general geology of the granitic region of southwestern California, in which these remarkable developments of minerals have recently been made, has been repeatedly described, but never very fully or exactly determined. At the time of the second geological survey, under Prof. J. D. Whitney, 1860 to 1865, the whole region was little known or settled, save at a few points on the coast. Professor Whitney gives certain broad outlines, but not based upon any detailed examination of that part of the State as a whole. He points out the disappearance of the Great Valley, with its clear definition of the two mountain systems on either side, the Sierra and the Coast Ranges, and the apparent intermingling of the two, as the Sierra bends toward the southwest, cutting off the central valley. This southwestward extension of the Sierra passes across what is now Kern County, and, with the Tehachapi Mountains, divides it into two well-marked portions. Professor Whitney observes that it is only by the age and position of the sedimentary rocks, and not by any topographical features, that the mountains below this line can be judged as to their relations. As the two systems approach and intermingle, the lines of disturbance are so closely related, and so influ-

---

*Letter from Prof. T. C. Hopkins, dated January 11, 1905.
† W. T. Schaller, Spodumene from San Diego County, Cal.; Bull. Dept. Geol. Univ. Cal., Vol. III, Sept., 1903, pp. 265–275.

enced by secondary ones, that the topography gives no clew to the actual facts of structure.*

His determination was that the Coast Range ends at about San Luis Rey, and is not traceable farther south; and that the principal mountains of what are now San Diego and Riverside counties belong geologically and orographically to the Sierra Nevada, which stretches on southward to form the long peninsula of Lower California.

This close relation of the region under consideration to the system of the Sierra was recognized and confirmed ten years later by Prof. W. A. Goodyear, who made a reconnaissance of the country in 1872. His notes were published many years afterward, in the eighth report of the State Mining Bureau, 1888. They are clear and vivid in their topographic portrayal. He describes† the geological and geographical features of San Diego County (including also what is now Riverside County) substantially as follows:

The broad mountain range which, stretching south from the San Bernardino Valley, occupies all the western part of San Diego County, from the Pacific east to the Colorado Desert, and south to the Mexican line, is essentially a region of granite, nine tenths of it being composed of this rock. Along the coast it is flanked by Tertiary sediments; and much of it, especially in the eastern part, is traversed by belts of highly metamorphic schists, micaceous and hornblendic, with a prevailing northwesterly course. The detailed topography is exceedingly complex, the ridges trending in different directions and inclosing valleys at various altitudes.

Along the shore extends the Tertiary mesa, which rises gently eastward, and sometimes reaches far inland among the mountains, to heights varying from 500 to 800 feet. The mountains gradually increase in elevation, until at some 50 miles from the coast they form a crest line of some 6000 or 7000 feet, from which they fall off steeply and rapidly on the eastern side some 5000 feet to the western edge of the Colorado desert; thus resembling, says Professor Goodyear, on a smaller scale, the form and contour of the true Sierra Nevada.†

Such is the general character until the northern part of San Diego (now Riverside) is reached, when a more east and west trend appears, culminating in the grand peak of San Jacinto, near the San Bernardino line, whose height is given by Lieutenant Wheeler as about 11,000 feet.

Professor Goodyear then describes in some detail his examination of the rocks and the structure in ascending the cañon of the San Diego and traversing the ridges and valleys to Ramona, Ballena, Santa Ysabel, Julian, and Banner, some of which are in the region of the recent discoveries. Everywhere the predominant rock is granite, of

---

*Geol. Survey of Cal., Vol. I, 1865, pp. 167, 168.
† California State Mining Bureau, Rept. VIII, 1888, pp. 516–522.

varied type, sometimes becoming a syenite, "consisting of feldspar and hornblende, with but little quartz and almost no mica, which might almost be called a diorite." This latter rock is now recognized as a distinct formation from the granite, as elsewhere described herein. He notes the occurrence of the orbicular diorite at several places, saying that the granite "often contains dark-colored hornblendic nodules, * * * whose texture is still granitoid." The pegmatite dikes, now found to be so rich in gem-minerals, are well described, as follows: "The granite country here (*i. e.*, near Julian) is frequently traversed by veins of very coarse granite, which sometimes furnish plates of mica one to two inches in diameter, with correspondingly large blocks of feldspar"; he also noted the black tourmaline crystals in the same veins, but did not encounter the colored ones.

The ascent of Cuyamaca Mountain is then described; and the general features of the whole region, as seen from that fine point of observation, are so clearly presented that it is well to quote them in detail. The view is extensive and grand—northward to the San Jacinto and San Bernardino mountains, westward along the coast and out to sea, and southward far into Mexico; northeast lies the Colorado Desert and the Coahuila Valley, and a long stretch of the San Bernardino Mountains running toward the Colorado River along the northeast side of the desert.

"From this standpoint," Professor Goodyear says, "the whole country from just back of San Diego * * * to the western edge of the desert is like an angry ocean of knobby peaks, more or less isolated, with short ridges running in every possible direction, and inclosing between and among them numerous small and irregular valleys. As a general rule, the higher peaks and ridges rise from 1000 to 2500 feet above the little valleys and cañons around their immediate bases. But in going eastward from the coast, each successive little valley is higher than the one * * * preceding, and the dominant peaks and ridges also rise higher and higher * * * until we reach the irregular line of the main summit crest, or water-divide, * * * when the mountains break suddenly off and fall within a very few miles from 4000 to 5000 feet or more, with an abrupt and precipitous front to the east, to the western edge of the desert. It thus follows that this chain of mountains, as already stated, though made up of a confused mass of minor ridges and peaks of granite, having in their detailed topography but little connection with or relation to each other, nevertheless has a general orographic form very closely allied to that of the Sierra Nevada in the more central portions of the State."

In the volume published by the Miners' Association, in 1899, for the California meeting of the American Institute of Mining Engineers, the following similar, though much more recent, account is given as to this region. Referring to the mineral wealth of southern California, that term is defined as that portion of the State south of the Tehachapi Mountains—" which unite the Sierra Nevada and Coast Ranges, inclosing the upper (*i. e.*, southern) end of the central valley, and topographically dividing the State into two regions of distinct characters." The belt on the west is fertile, salubrious, rich, and prosperous, the mineral resources are chiefly oil and asphalt; otherwise, the mineral wealth of southern California lies eastward, in a very different area, comprising most of San Diego, Riverside, San Bernardino, and Los Angeles counties. The chief mining region is the desert country east of the Coast Ranges, " a region of rugged mountains, bare and forbidding hills, and sandy plains, divided by a series of mountain elevations into the Mojave and Colorado deserts." Gold and other mines exist all through this arid country, in which the rocks present great diversity, of all ages and types, the igneous ones being very numerous and marked. This report was chiefly concerned with gold mining; but the general presentation given is very good, and may well be taken here as descriptive of the recently developed gem-regions of southern California.

The two contrasted areas above noted have their separate types of gem production—those already mentioned as III and IV—in the extreme southern part of the State. The fertile region on the west is the Tertiary plateau or coast mesa; then comes the broad zone of granitic hills and ridges, so vividly pictured by Professor Goodyear, and identified by Professor Whitney, as in reality the southwestern extension of the Sierra, although in its geographical relations it occupies the place of the Coast Range. Through this hilly zone run the pegmatite veins, so rich in lithia minerals—tourmaline, kunzite, lepidolite, etc.—and in garnet, beryl, and topaz, at many points from the Mexican border to the heights of the San Jacinto Mountains. Then comes, on the east, the steep falling-off of the mountain area, as described by Professor Goodyear, to the arid stretch of the Colorado Desert, bounded on the northeast by the San Bernardino range, beyond which, to the north, lies the Mojave Desert, with its borax mines. In the Colorado Desert, among the volcanic rocks, are the opal and turquoise localities mentioned under III—some of the latter far up on the San Bernardino heights, and others away to the eastward in Nevada and Arizona.

It is with the western division—that of the granitic hill country—that the present report has principally to deal, although the other gem-producing areas are also described.

HISTORICAL OUTLINE.

The first discovery of colored gem-tourmaline in the State goes back as far as 1872, when Mr. Henry Hamilton, in June of that year, obtained and recognized this mineral in Riverside County, on the southeast slope of Thomas Mountain. These colored tourmalines, now found at a number of points, were not encountered by Professor Goodyear, who particularly noted the black tourmalines in the pegmatite veins, in his geological tour through San Diego County, in the same year, referred to above; but his reconnaissance was a little south of the gem-tourmaline belt. Some mining was done at this point, and fine gems were obtained. In the course of years, three localities were opened and more or less worked in this vicinity; so that in the author's report on American gem-production for 1893, the following statement appeared:*

"Tourmalines are mined at the California gem mine, the San Jacinto gem mine, and the Columbian gem mine, near Riverside, California. These three mining claims cover the ground on which the tourmaline is found, and are situated in the San Jacinto range of mountains in Riverside County, California, at an altitude of 6500 feet, overlooking Hemet Valley and the Coahuila Valley, and 27 miles from the railroad. The formation in which the crystals are found is a vein from 40 to 50 feet wide running almost north and south through the old crystalline rocks which make up the mountain range. The vein in some places consists of pure feldspar, or else feldspar with quartz, in others all mica, and in others rose-quartz and smoky quartz. The tourmalines vary in size from almost micrograins to crystals 4 inches in diameter. They are most plentiful in the feldspar, but are found in other portions of the vein, sometimes in pockets and sometimes isolated. The larger crystals generally have a green exterior and are red or pink in the center. Some of the crystals contain green, red, pink, black, and intermediate colors; others again are all of uniform tint—red, pink, colorless, or blue. Associated with the tourmalines are rose-quartz, smoky quartz, asteriated quartz, and fluorite, and some of the quartz was penetrated with fine, hair-like crystals of tourmaline, strikingly like a similar occurrence of rutile."

It may seem remarkable that this locality of gem-tourmalines should have been unrecorded in the earlier lists of California minerals given by such authorities as Professor Blake and Mr. Hanks in the reports of the State Mining Bureau for 1882 and 1884. But the parties who knew of the occurrence did not make it public for some years, and the earlier specimens were taken out quietly and their locality not divulged. The writer had positive knowledge as to the facts, however, and possesses a fine specimen obtained prior to 1873.

*George F. Kunz: Min. Res. U. S., Rept. U. S. Geol. Survey, 1893, p. 18 (reprint).

The second important discovery in this region was made, or at least announced, twenty years later, in 1892, by Mr. C. R. Orcutt—the great locality of lithia minerals at Pala. Some allusions to red tourmaline from uncertain sources in this part of the State had appeared before; but nothing very specific. In the list of California minerals prepared by Prof. William P. Blake in 1880–82,[*] and also quoted in that of

ILL. No. 6. "BRIDAL CHAMBER," LEPIDOLITE MINE, PALA,
SAN DIEGO COUNTY—SHOWING TOURMALINE
CRYSTALS EMBEDDED IN THE ROCK.

Mr. Henry G. Hanks, published in 1884,[†] references are made to the recent discovery of rubellite, for the first time in the State, associated with lepidolite, "in the San Bernardino range, southern California." The general description is precisely that of the Pala specimens, but the location is very indefinite. Mr. Hanks refers to the same association under lepidolite, and mentions a specimen in the State Mining

[*] State Mineralogist, 2d Rept., 1880–82, p. 207, Appendix.
[†] Ibid., 4th Rept., 1884, p. 389.

Bureau, from San Diego County, and remarks that "this may at some future time be found profitable to extract lithium from it"*—a prediction abundantly verified now. Mr. Orcutt, however, was the first to make the locality known. It was noted by the author in his report for 1893, where the following account was given: †

"Mr. Charles Russell Orcutt has announced a new and remarkable occurrence of pink tourmaline in lepidolite, similar to that of Rumford, Maine, 12 miles south of Temecula, near San Luis Rey River, in San Diego County, the southern county of California, and it has already become celebrated from the abundance and beauty of the specimens yielded, as much as twenty tons having been sent East for sale. Through San Diego County runs the Peninsula range, rising several thousand feet between the coast and the Colorado Desert. In these granite mountains are diorite intrusions and some metamorphic schists, etc. West of the summit lies a parallel belt of granitic rock characterized by dikes of pegmatite, in one of the largest of which occurs this great deposit of lepidolite with tourmaline. In Pala, a little west of Smith's Mountain, in the Peninsula range, * * * a ledge of lepidolite containing rubellite has been traced for over half a mile. It consists of a coarse granite, penetrating a norite rock, and including masses of pegmatite. Small garnets occur in the granite, and black tourmaline, with a little green tourmaline. The lepidolite appears in the southern portion, finally forming a definite vein which at one point is twenty yards wide. The rubellite is chiefly in clusters and radiations, several inches in diameter, also occasionally as single crystals, and the specimens of deep pink tourmaline in the pale lilac mica are remarkably elegant. About eighteen tons were mined during 1892."

The next important discovery was made six years later, in 1898; this was the wonderful Mesa Grande locality, some 20 miles southeast of Pala. There are various stories about the Indians having known it for many years, and the most familiar account is that given further on under Tourmaline. But the fact that some of the highly colored crystals are found in Indian graves in the vicinity, suggests that they may have been known and valued perhaps for a very long time. The ledge in which they occur is exposed by erosion on the side of the mountain; and the natives had certainly learned where to find crystals, and had them in their possession for some years before the whites knew anything about them. It is even said that they had learned how to do a little rude blasting, and thus to reach the cavities in which the minerals occur. It was not until 1898, however, that this now famous locality was made known to the world.

---

*State Mineralogist, 4th Rept., 1884, p. 254.
† Rept. U. S. Geol. Survey, 1893, Min. Res. U. S., pp. 17, 18 (reprint).

The discovery was announced in the author's report for 1900, on the production of precious stones in the United States, as follows:[*]

"In 1898, while prospecting in Mesa Grande Mountain, San Diego County, California, for lepidolite, a large ledge was observed that appeared to be a mass of this mineral. This locality is at an altitude of 5000 feet on the Mesa Grande Mountain, a region in which no geological work had up to that time been done. The first few blasts showed that lepidolite was present in quantity, and also in larger and more brilliant scales than in the well-known locality at Pala, Cal. Both in the lepidolite and in the associated quartz there are magnificent crystals of tourmaline, and, as at Pala, the rubellite variety predominates. The new locality differs, however, in having the tourmaline in distinct, isolated crystals. Many of these are translucent, or even transparent, and occur as large, separate crystals, with perfect prisms and terminations. They differ in both these respects from the Pala crystals, which are nearly opaque and grouped in radiations almost blending into the matrix, which latter is lepidolite, with rarely ever any quartzite. The rubellite seems the predominating variety at Mesa Grande Mountain; but there is also a large proportion of parti-colored crystals—i. e., those made up of three, four, or five distinct sections, as at Haddam Neck, Conn., and Paris, Me.; others present the Brazilian type, in which several different colored tourmalines appear, as though included one within the other. In the Brazilian crystals, however, the interior is generally red, inclosed in white, and the exterior green. This concentric arrangement is reversed in the crystals from Mesa Grande Mountain, which are generally green in the interior, or yellow-green, inclosed in white, with the exterior red. The habit of the crystals is also very interesting, in that many of them, when doubly terminated, end in a flat, basal form of pyramid, and are not hemimorphic, as tourmalines generally are."

For several years, these above noted were the only gem mines of this region, and their product was highly esteemed. But in 1902 began a succession of new discoveries that have attracted great attention. On Pala Chief Mountain and on Heriart Mountain began to be found not only fine-colored tourmalines, but the novel and remarkable gem-spodumene, designated as kunzite. This last-named mineral was found by Mr. Frederick M. Sickler, at what is now known as the White Queen mine, on Heriart Mountain, east of Pala, early in 1902; it is claimed, indeed, that he had obtained one or two pieces some time before, but it was not identified. In July, 1902, Mr. Sickler visited San Diego and Los Angeles, and showed specimens to local jewelers and collectors, none of whom recognized it. The first determination was made by the writer, from specimens sent by Mr. Sickler early in 1903.

---

[*] Min. Res. U. S., Rept. U. S. Geol. Survey, 1900, p. 33 (reprint).

The great Pala Chief mine, which has given its name to the middle . one of the three ridges or mountains at Pala, and has yielded magnificent tourmalines and the largest and finest gem-spodumene crystals, was located in May, 1903, by Frank A. Salmons, John Giddens, Pedro Peiletch, and Bernardo Heriart. The actual discoverers were probably the two last named, the Basque prospectors who had already been working and locating claims with the two Sicklers, father and son, on Heriart Mountain, the ridge a little to the east. Mr. Salmons has been the principal operator, however, of this very notable mine.

The first public announcement of these discoveries appeared in the writer's report on gem-production in the United States for 1902, having been introduced late, while the report was being printed in 1903.* They were also described by the writer in "Science" for August 28, 1903, and in the American Journal of Science for September of the same year.†

Meanwhile, on September 8, 1902, gem-tourmaline had been discovered on Aguanga Mountain, some 5 miles south of Oak Grove, by Mr. Bert Simmons. This locality lies nearly east from Pala and south from that at Coahuila, next to be mentioned, and about equally distant from the two, some 15 miles. Kunzite has since been found on the same claim.

On May 30, 1903, Mr. Simmons discovered both colored tourmalines and kunzite in Riverside County, some 10 miles west of the old Hamilton (first) discovery. The locality is on Coahuila Mountain, about 20 miles northeast of Pala. The mine was for some time known as the Simmons mine, but has been sold to Mr. E. A. Fano, of San Diego, and is now called by his name. This is one of the most promising and productive mines of the region.

The discoveries at and around Ramona followed in rapid succession, in 1903. Some had been made several years earlier, but they had not attracted much notice. Essonite garnet was reported near Ramona in 1892, by D. C. Collier, and also fine epidote. Much of the essonite found hereabout is of rich color and fine gem quality.

Several mines, with this "hyacinth" variety of garnet and more or less of beryl and tourmaline, were located in May, July, and September, 1903.

On October 3d of that year, topaz was discovered in the same vicinity, by James W. Booth and John D. Farley. This was a novel and important addition to the gem products of the State. The crystals are of various sizes, some of them large, often transparent, and range from colorless to pale shades of blue, much resembling those from the old and well-known locality at Sarapulka in the Ural Mountains.

These minerals will be described further on, in the body of this report,

---

*Min. Res. U. S., Rept. U. S. Geol. Survey, 1902, pp. 848, 849.

†Science. Vol. XVIII (new ser.), No. 452, 1903, p. 280; and Am. J. Sci. (4), Vol. XVI, 1903, pp. 264-267.

and the several mines will be enumerated, with their special products, in the section following.

As was noted before, the garnet and topaz belt seems to run on a distinct and parallel line somewhat southwest of the tourmaline-kunzite mines. The main localities are near Ramona; but if a line be drawn from that point southeast to the Mexican border. it will strike another great garnet region near Jacumba Hot Springs. These localities have only recently been much known or examined. They were first described in the writer's report on gem production for 1903, together with the Ramona discoveries above noted,* as follows:

"Essonite has been found at a number of localities in deposits spread over a considerable territory from 9 to 10 miles northeast of Jacumba Hot Springs, San Diego County, Cal., usually associated with granite and granular limestone. At three of the places some gem material has been found. Associated with it is a little vesuvianite and crystallized quartz. Eleven localities in this region are noted. Essonite has also been found near San Vicente, El Cajon Mountains, but the crystals were full of imperfections. The finest essonite crystals are obtained at Ramona, San Diego County, associated with green tourmaline, white topaz, and beryl, occasionally in perfect dodecahedrons and trapezohedrons, of rich yellow to orange-red color, and very brilliant. They have also been discovered at Warner's Ranch, Mesa Grande, Santa Ysabel, Gravilla, and Julian, San Diego County; Deer Park, Placer County; Laguna Mountains and Jacumba, and also at several places below the Mexican line. As some of the crystals were of exceptional brilliancy, it is possible that on further development many fine gems will be obtained."

The name Jacumba is used in a very general way for any place within a few miles of the store and springs. It properly belongs to a small valley surrounded by mountains of granite, and locally noted for its earthquakes and hot springs, situated close to the Mexican line. The springs are liable to great fluctuations of level, and there are extensive lava-flows among the mountains around, so that the region appears to be one of recent volcanic activity. As yet, however, it has not been accurately mapped or geologically examined. The springs are both hot and cold, variously impregnated with mineral substances, and are likely to become important as a health resort, especially as the country still abounds with wild game. They are situated on the projected railroad line skirting the frontier, from San Diego to Phœnix, Arizona—74 miles east of San Diego and some 20 miles from Campo, in the S. E. ¼ of Sec. 12, T. 18 S., R. 7 E., S. B. M. A short distance east is the main mountain crest, and then a steep descent to the Colorado Desert.

Throughout this region around Jacumba, essonite garnet is found at

---

* Min. Res. U. S., Rept. U. S. Geol. Survey, 1903, p. 19 (reprint).

various points, together with black tourmaline and some beryl. As elsewhere in all the granite country of San Diego County, these minerals are associated with pegmatite veins, though at one or two points the garnets are reported in a limestone. The mines best known are situated in the Santa Rosa Mountains, several miles northeast of Jacumba. Not much working has been done as yet, but there is likely to be a good deal more soon. One mine, the Dos Cabezas, in which the garnets occur in a marble, has been known for some ten years, and occasionally worked, yielding many fine hyacinths.

The country hereabout is very wild, rugged, and inaccessible, and wood and water are scarce. If the railroad is opened through, this may become an important region of gem-production.

In the whole hilly country of the granite and diorite, west and south from these lines of opening, here briefly indicated, constant reports are coming in of interesting mineral discoveries. The orbicular diorite, or napoleonite, elsewhere described, near Dehesa, and the newly discovered lilac dumortierite, not far from the same place, may both become valuable ornamental stones, if procurable in quantities sufficient for such purposes. These are described in the body of this report. The whole country seems full of possibilities for precious and semi-precious minerals; and years must yet pass before it will be so fully explored that any complete estimate of its resources can be formed. Meanwhile this report brings together most of what has been discovered, and also of what has been done thus far, in regard to the gem-minerals of southern California.

RECOGNITION.

In issuing this report, the writer desires to bear testimony to the work of those who have preceded him in the study of California mineralogy, and also of those who have directly aided or contributed to the gathering of the facts herein presented. A very brief review may be given in the first place of the history of mineralogical investigation in the State of California, in order to link the present with the past and to show something of the course of development.

Within four years after the first discovery of gold, in 1847, at Sutter's mill, El Dorado County, a geological survey of the State was organized, with Dr. John B. Trask as its director; his preliminary report was made in 1851, and four annual reports were issued in the years 1853 to 1856 inclusive. These dealt principally with the Sierra Nevada and Coast Ranges, with especial reference to gold. Work was then suspended for several years, until the organization of the Second Geological Survey, in 1861, under Prof. J. D. Whitney. This was begun and carried out on a noble and comprehensive scale, until the unfortunate stoppage of appropriations for it in 1874. The work of

Professor Whitney and the eminent scientific experts associated with him was thus suddenly broken off. Portions of the work, that had been done and were ready for publication, were subsequently issued elsewhere under the auspices of scientific societies.

After six years, the State Mining Bureau was organized by the Legislature in 1880, and has been maintained from that time, with a great amount of excellent and valuable work by many able men. The office of State Mineralogist was created, and Mr. Henry G. Hanks was appointed; he held the position from 1880 to 1886, and gave a very important impetus to mining and mineralogical interests. His reports are full of valuable material. He was followed in succession by William Irelan, Jr., 1886–1893; J. J. Crawford, 1893–1897; A. S. Cooper, 1897–1901; and the present State Mineralogist, Lewis E. Aubury, from that time on. The Bureau now occupies a large and impressive building in San Francisco, and has gathered extensive collections of the minerals of the State, as well as of general mineralogy for comparative study.

Among those who have dealt particularly with the regions, or the topics, considered in the present report, in distinction from the mining of gold and other metallic productions, which have naturally held the first place in the work that has been done, reference may be made to the following persons: The earliest list of minerals of the State was prepared by Prof. William P. Blake, and published in the second report of the State Mining Bureau, 1882 (appendix); this was followed by other lists based upon it and adding to it, by Mr. Henry G. Hanks, published in his fourth and sixth reports, 1884 and 1886. These lists are quite full, and contain many valuable notes on the peculiarities of the minerals at the localities described.

With special regard to the finding of diamonds in the State, numerous articles have appeared, from an early date, in the American Journal of Science, the proceedings of the California Academy of Sciences, the Mining and Scientific Press, and other publications of various kinds; these will be referred to in the section upon diamond occurrence, later. The Rev. C. S. Lyman was the first to describe a California diamond, in 1848; and Prof. Benjamin Silliman published several early accounts. In 1854, Dr. Melville Attwood, an important contributor to the geological study of the State, and author of valuable papers published by the State Mining Bureau, called public attention to the occasional presence of diamonds in the gold gravels, and the possibility of further discoveries. In 1871, Prof. W. A. Goodyear published a reference to the subject in Prof. R. W. Raymond's volume on "Mineral Resources West of the Rocky Mountains." A full account of all the discoveries up to date was given by Mr. Hanks in his second report of the State Mining Bureau for 1882, with comparisons between the California occurrences and those in other parts of the world; and this was supplemented in his reports for 1884 and 1886. The last and most comprehensive article on this subject was that of Mr. J. W. Turner, in 1899 ("Diamonds in California"), published in the American Geologist, Vol. XXIII, in which all the occurrences are given, up to that time.

The geology of the granitic region of the southwestern section of the State, spoken of formerly as San Diego County, but including also what is now the County of Riverside—the region which is principally treated of in this report, on account of its recent very notable development as a gem country—has attracted more or less attention from an early period, and has been described partially and briefly by a number of observers. Prof. J. D. Whitney, as will be shown further on, recognized some of its leading geographical features in his first volume on the Geology of California, published in 1865. Prof. Rossiter W. Raymond, in his report for 1872, on "Statistics of Mines and Mining West of the Rocky Mountains," devotes a chapter to San Diego County and its early gold developments, based largely on the studies of Mr. C. A. Lockhardt, who examined and reported upon it in 1870. Prof. W. A. Goodyear, who is cited herein, later, traveled over portions of the region in 1872, and described it very vividly and clearly, although his observations did not appear for several years. Mr. H. G. Hanks, in his sixth report, 1886, gave an account of San Diego County and its mineral resources, with a large and valuable map. Professor Goodyear's observations, above noted, appeared in the ninth report, issued by Mr. Irelan, in 1889.

In the Proceedings of the California Academy of Sciences for 1888, Mr. Waldemar Lindgren published a series of "Notes on the Geology of Baja California" (2d series, Vol. I, p. 173), with a profile from San Diego to the Colorado Desert. A number of papers and articles on this region have appeared since 1890, among which the following may be mentioned: Relation of the metamorphic and granitic rocks of the Sierra Nevada and Coast Ranges; H. W. Fairbanks; Amer. Geologist, Vol. XI, 1893, p. 69;—Geological sketch of Lower California; S. F. Emmons and G. P. Merrill; Bull. Geol. Society of America, Vol. V, 1894, p. 489;—Geology of San Diego County; H. W. Fairbanks; West American Scientist, Vol. X, 1901, p. 86; –and several articles by Mr. C. R. Orcutt on Mines and Minerals of San Diego County (the same, Vol. III, p. 69) and on the Colorado Desert (the same, Vol. VII, 1890, p. 55, and Vol. XII, 1901, p. 102), and previously in the Tenth Report of the State Mining Bureau, 1890.

These and all other reports, articles, and publications of whatever kind on California geology, may be found recorded in accurate detail, and often with valuable notes on any special features, in the admirably careful and systematic bibliography of the subject prepared for the

State Mining Bureau by Brigadier-General Anthony W. Vogdes, U. S. Engineers, and published by the State Mining Bureau as Bulletin No. 30.

It remains to speak more particularly of those who have personally contributed, directly or indirectly, to the data concerning gem-minerals especially dealt with in this report, whether as workers in the field of actual discovery and development, or as collectors and students, or as correspondents furnishing valuable facts. In many cases, they have combined some or all of these relations, and it is a pleasure to recognize their services in any of these departments.

Mr. Melville Attwood, F.G.S., a careful worker, a microscopist as well as a mineralogist, first really called the attention of the world to the finding of diamonds in the hydraulic gold washings of California, in 1854. He is said to have been the first also to have identified the silver ores in the great Comstock lode. He was a mining geologist of recognized ability, and contributed important papers to the reports of the State Mining Bureau, especially that on the milling of gold quartz (Second Report, 1882) and on the lithology of wall rocks (Eighth Report, 1888).

Mr. Henry G. Hanks, the first State Mineralogist of California, had much to do with the development of many of the mineral localities, and his admirable reports and scientific papers kept the world well informed of the progress of mineralogy and mining in the great Golden State. His lists of minerals have been already referred to; while, perhaps, his most prominent special work was that on the borax deposits, published in the third report of the State Mining Bureau (1883).

Under his successors, the scientific work of the Bureau has been worthily carried on, and the collections increased to their present noble extent.

Among the earlier students and writers, the names of Prof. William P. Blake and Prof. Benjamin Silliman, Jr., are not to be overlooked. The former prepared the earliest list of rare California minerals, for the second report of the State Mining Bureau, in 1882, besides numerous articles in scientific journals; and the latter was associated with Professor Whitney and Mr. Hanks, as far back as 1867, in the celebrated presentation and discussion of diamonds from the gold gravels, before the California Academy of Sciences.

Since 1878, Mr. Charles Russell Orcutt, of San Diego, Cal., editor of the West Side Scientist—a modest publication that has yet done more than any other concerning the flora of Lower California—from time to time made excursions for plants, especially cacti, and during these trips collected mineral specimens which he sent to the author for determination. His was the first information received of the great lepidolite mine at Pala, and several of the other gem-producing localities in southern California. He has also published important papers in the reports of the State Mining Bureau, especially on the mineralogy and geology of the Colorado Desert (Tenth Report, 1890) and in the Mining and Scientific Press for the same year (Vol. VII).

Mr. Henry S. Durden, for many years curator of the collection of the State Mining Bureau, has repeatedly sent the writer information concerning the occurrence of precious stones in California, using great care and discrimination in transmitting such announcements.

Mr. Max Braverman, an ardent and careful collector of minerals, residing at Visalia, Cal., has for a long time been contributing valuable information concerning the finding of the topazolite, chrysoprase, hyalite, and various other minerals of his own vicinity and region; and recently, with a public spirit and generosity worthy of the highest citizenship, he has presented his collection, the work of many years, to the Golden Gate Museum at San Francisco, when he had received offers for its purchase from institutions in other States.

Mr. Dwight Whiting, formerly of Boston, has long been interested in securing information about gem localities in California, and since 1893 has furnished many new and valuable facts, which have been recorded in the writer's annual reports on the Production of Precious Stones.

With regard particularly to the remarkable discoveries of gem-minerals in the last few years, in San Diego County :—

Mr. Fred M. Sickler and his father, M. M. Sickler, of Pala, have for many years been interested in the subject of mineral development in southern California, and it was Fred M. Sickler who first sent to the writer a mineral which the California lapidaries did not recognize, and had been unable to cut, owing to a peculiar cleavage. This mineral, when it reached New York, was identified by the author as a form of spodumene, and was subsequently given the new name of *kunzite* by Dr. Charles Baskerville. The Messrs. Sickler have, since that time, paid much attention to the development of mining properties in the Pala region, have located a number of claims, and have contributed many descriptive letters, which have materially aided in the preparation of this State report.

Two Basque Frenchmen, Bernardo Heriart (after whom Heriart Mountain is named) and Pedro Peiletch, have been most careful prospectors, and have assisted in the locating of a number of claims, in the same region, of kunzite and other gem-minerals, including the great Pala Chief mine, of which they were apparently the first discoverers.

It was Frank A. Salmons, now County Clerk of San Diego County, who thoroughly developed the Pala Chief mine, in which have been found the greatest deposits of the gem-spodumene and rubellite in that vicinity. To him is due the credit of having sent the finest specimens of these minerals from California that had yet been seen.

Dr. W. T. Schaller, a graduate of the University of California, visited the gem regions of San Diego County in the summer of 1903, during the preparation of his thesis on spodumene, which was published by the University of California. He also visited the State in June and July of 1904, in behalf of the Department of Mining Statistics of the U. S. Geological Survey, to study the deposits of lithia minerals for a bulletin to be issued by the Survey. The results of this investigation will be published later.

The writer himself went to California in 1890, in the interest of the Eleventh United States Census, and while there visited a number of the localities and local collections. He also published all the information that could be obtained upon the subject of California precious stones in his volume on "Gems and Precious Stones of North America," issued in New York in 1889, and the two appendices to it in 1890 and 1892; and also annually in the reports of the Department of Mining Statistics of the U. S. Geological Survey, which have appeared from 1882 up to the current year (1905).

With special reference to this report on California gems herewith presented to the State Mineralogist, Mr. Aubury, Mr. W. H. Trenchard, of San Diego, has lately visited nearly all the localities of precious stones in the southern portion of the State, making measurements, obtaining facts, collecting specimens, and securing photographs, many of which are printed herein.

To Mr. Samuel G. Ingle, of San Diego, the author feels under great obligation for information, specimens and photographs, which he has sent from time to time, and which have assisted in presenting many of the facts contained herein; also to Mr. H. C. Gordon, who, as a very careful and observant correspondent, has furnished numerous data, together with many specimens of the gem-minerals and their associations, as well as photographs of localities, several of which are reproduced in the following pages.

Thanks are due to the Hon. L. E. Aubury, at whose suggestion this precious-stone report was made, and under whose direction the list of minerals in the State Mining Bureau was prepared, for his continued courtesy in furnishing information, photographs, and assistance throughout the entire preparation of this work.

In regard to the future of precious-stone mining in southern California, although the great diamond output of to-day is not adequate for the world's demand, it will require possibly a change of fashion or new adaptability of materials to consume all the semi-precious stones that are likely to develop if mining is continued in this line with much more energy. California as a tourist's resort has the advantage over many other places in being an attractive country, drawing many thousands of strangers, who are generally more or less affluent; and if the precious stones themselves are handsomely or quaintly cut, and are in all instances what they are represented to be, there is likely to be a large demand created in this way. Unfortunately, at other places in the United States, foreign or artificial minerals have been substituted for native material. If the smoky quartz, lepidolite, and like minerals were worked up into desk weights, seals, charms, etc., a great quantity could also be sold not only in California, but elsewhere. It was the development of an industry like this in the Ural Mountains of Russia, brought about by Catherine II. sending two lapidaries to that region, that led to the employment of fully one thousand people in the Ural district. As the chrysoprase, the turquoise, and the tourmaline of California, when not of fine gem quality, have been cut into crude beads of East Indian type, into small forms of mosaic work, and the like, a large quantity of this material has found a definite market in Europe as well as in the United States, that otherwise would have been a loss in mining. The development of proper lapidary interests of this kind will surely do much to sustain the mining of gems in California and add appreciably to the wealth of the gem regions.

THE PROPERTIES OF GEMS.

It is difficult to define to-day what is meant by "a precious stone," for the mineralogist would give one definition, the jeweler another, the archæologist still a different one, while the scientific collector and the gatherer of curios and souvenirs would not agree with any of the others. A gem-mineral or a "precious stone" may be defined as a mineral of any sort, distinguished for its beauty, durability, or rarity, especially when cut and polished. There are only a few really precious stones: the diamond, the ruby and sapphire (identical in composition), the emerald, and occasionally the pearl (which is of animal origin) is included; formerly also the opal.

Some twenty years ago jewelers sold only a few varieties of stones; to-day they keep in stock anything known to the mineralogist and demanded by the public. The consumption of gems is larger than most people realize. Take one trade alone, for example. The watches manufactured annually in the United States use from seven to twenty-one jewels for each watch. The consumption annually amounts to over five million ruby and sapphire watch-jewels, and over seven million garnet jewels; while over 15,000 carats of bort diamonds are consumed in cutting these jewels.

In looking for gems, the prospector should be provided with a few specimens representing the scale of hardness, and have the means in camp to determine the specific gravity. In this way he can separate the positively worthless from the possibly valuable, even though he can not fully identify the minerals.

3—MB

A large number of the many varieties of precious and semi-precious stones and rare minerals are found in California, and systematic search will increase the production until California takes high rank as a gem State. In 1901, California produced quartz crystal to the value of $17,500; tourmaline to the value of $20,000; and turquoise, over $20,000; the lepidolite amounted to $27,500; gold quartz, etc., over $50,000; mother-of-pearl and pearls, over $15,000; and souvenir material, probaably over $20,000. The grand total amounted to nearly $175,000. In 1903 kunzite was produced to a value of $20,000; tourmaline, $20,000; chrysoprase, $15,000; turquoise, $40,000.

Color.—The color of many gems is variable; the sapphire blue, the ruby red—both varieties of the one species of corundum. The garnet is popularly supposed to be a blood-red or purplish-red stone, but it varies through red of several shades to brown, black, green, yellow and nearly white; the tourmaline, green, red, pink, yellow, white; the topaz, yellow, white, and blue.

Diaphaneity.—The ability to transmit light affects materially both the beauty and the value. A stone is transparent when the outline of an object is clearly seen through it; subtransparent, when the object may be seen, but the outline is indistinct; translucent, when light is transmitted, but objects are not seen; subtranslucent, when merely the edges are translucent; opaque, when no light is transmitted.

Luster.—Luster is the manner of reflecting light. It is described as metallic luster, or the brilliant appearance of polished metal; adamantine, the luster of the diamond; vitreous, the luster of glass; resinous, like the surface of pine resin; waxy, like beeswax; greasy, like a freshly oiled surface; pearly, like mother-of-pearl; silky, having the sheen of silk.

Refraction.—The apparent breaking of a spoon when immersed in a tumbler of water is a familiar illustration of the bending back or refraction of light. A line seen through Iceland spar appears double; diamond, garnet, and all minerals crystallizing in the isometric system are single refracting. All minerals belonging to other systems of crystallization, like the ruby and topaz, are double refracting.

Dispersion.—When a ray of light passes through a prism of flint glass, it gives the spectrum or rainbow band. Refracted rays of white light may be decomposed into several rays differently colored. This is called dispersion, and gives "fire" to gems, notably in diamonds and zircons.

Pleochroism.—Double refracting minerals show a variation in color by transmitted light, when viewed in different directions, due to the differential absorption of the white light. Tourmaline, andalusite, iolite, chrysoberyl, and epidote are good examples.

Fluorescence and Phosphorescence.—Some gems, when exposed to a powerful light, or when heated, will emit light. If this emission of light lasts only as long as the exciting agent is applied, it is called fluorescence; whereas, if the emission persists after the removal of the cause, it is called phosphorescence.

Hardness.—The degree of resistance to abrasion. This property is most essential to gems, if they are to receive and retain a high degree of polish and stand long use. Hardness is not the same as toughness. The diamond is very hard, but not tough—in fact, it is very brittle, easily broken by a blow.

Scale of Hardness.

1. Talc (lowest).	6. Orthoclase.
2. Gypsum.	7. Quartz.
3. Calcite.	8. Topaz.
4. Fluorite.	9. Corundum.
5. Apatite.	10. Diamond (highest).

Should the gem scratch (*note:* quartz will scratch feldspar) and be scratched by any unit of the scale, the hardness of the two is the same; should the gem scratch the one below, and be scratched by the one above, its hardness lies between that of the two test units.

Specific Gravity.—This is the density of the gem, compared with that of its own volume of distilled water at a temperature of 39° F. Weigh in the air, then in water; divide the weight in the air by the loss of weight in water, and the quotient is the specific gravity. For example, a gem weighs 4 grams in air, and only 3 when immersed in water; then the loss of weight is 1, and 4 divided by 1 gives the specific gravity of 4. The specific gravity test is one of the most definite means of identifying a gem.

Electricity.—Some gems become electrified by friction, and as a result attract or repel certain substances. This is especially noticeable in topaz, tourmaline, and amber.

Fusion.—Some gems will melt easily before the blowpipe, as some varieties of garnet, kunzite, etc.; while others are infusible, like quartz, topaz, etc.

Cleavage.—The tendency to break in a direction parallel to certain planes in the crystal, the resulting cleavage faces being smooth and often very brilliant.

Fracture.—When the mineral is broken in any other direction than that of the cleavage. The fracture may be conchoidal, uneven, or irregular, resembling a shell, even when the surface, though not a plane, approximates to one; or hackly, when the elevations are jagged.

Form.—The external form of gem-minerals may be:

1. *Crystallized*, in solids bound by plane surfaces, according to the following systems:

Isometric—cube, octahedron, dodecahedron;

Tetragonal—square prism, square octahedron;

Hexagonal—hexagonal prisms and pyramids, rhombohedrons;

Orthorhombic—right prism with rhombic base, rhombic octahedron;

Monoclinic—oblique prism with a rectangular base, and oblique octahedron;

Triclinic—doubly oblique prism, and doubly oblique octahedron or pyramid.

2. *Crystalline*, when the mass appears to be made up of closely compacted minute crystals, which may be arranged as columnar, lamellar, granular, globular, botryoidal, reniform, dendritic, etc., in varieties too numerous to be mentioned here.

3. *Amorphous*, showing neither external nor internal signs of crystallization and possessing no absolutely plane surfaces.

DIAMOND.

H.= 10.   G.= 3.52.   The hardest of all gems; the only combustible one; the most highly refractive, surpassing all others in its "fire"; crystallizing usually in octahedrons, or combinations of octahedron, cube, dodecahedron, and tetrahedron, the faces being commonly curved. Colors embrace nearly all of the prismatic hues; white, yellow, and brown are the most numerous, blue, green, pink, and red stones are rare. The perfectly white stones without a flaw, or those of decided tints of red, rose, green, or blue, are most highly prized. Fine cinnamon, salmon, brown, black, or yellow are much esteemed. They are sold by the carat; the international carat weighs 205 milligrams, or 3.168 + grains troy.

This stone is the purest, hardest, and most brilliant of all gems.

The occasional discovery of diamonds within the limits of the United States, although at widely separated points, and rarely of valuable quality or size, is yet a matter of very considerable interest.

The various points at which diamonds have been found may be grouped into four areas or regions, as follows:

1. The Pacific coast—a number of localities in California, chiefly along the western base of the Sierra Nevada; with these may perhaps be included provisionally, a few occurrences in Oregon, Idaho, and Montana.

2. The region of the great lakes—along an irregular line extending from western Wisconsin, across Michigan and Indiana to the vicinity of Cincinnati, Ohio—this line being essentially that of the terminal moraine of the later ice-sheet of Quaternary geological time.

3. A few localities in central Kentucky and eastern Tennessee, the relations of which are not entirely clear as to their geological connection.

4. The Atlantic coast—a number of points in North Carolina, and a few in Virginia, Georgia, and Alabama, lying in a general way along the eastern base of the Appalachian Mountains, some quite among them and others farther removed toward the east.

The occurrence of diamonds in California, both in the recent placer deposits and in the auriferous gravels of the ancient stream-beds now covered by lava, has been known for many years, and a number of localities are on record. No large diamonds have been found, at any time; and now that almost all the gold mining is carried on by means of stamp-mills, any that do occur are crushed into minute fragments, which are not infrequently found in the sluices and batteries, and furnish the evidence, and often all the evidence, of their continued occurrence.

Many notices have from time to time appeared, both in local newspapers and in scientific journals, of the finding of diamonds in California. After making due allowance for errors and unfounded rumors, their actual occurrence in certain localities is well established; but their number and size have not been such as to render the search for them profitable. The fact of their presence is highly interesting, and some of the specimens possess both elegance and value; but as a rule they are small and rare. In almost all cases they occur embedded in the auriferous gravels, and are thence washed out in the search for gold. These gold-bearing gravels of California present two types: first, as loose material in the valleys and bars of the modern streams; and, second, as great accumulations of gravel occupying the valleys of much larger ancient streams, and now covered with masses of lava or compact volcanic tufa. The sides of the Sierra Nevada are trenched with cross valleys running down into the great trough-like valley of central California, between the Sierra on the east and the Coast Range on the west. Along this great depression, the drainage from the mountains on both sides finds its way to the sea through the Sacramento and San Joaquin rivers, flowing respectively from the north and from the south into the Bay of San Francisco, where a break in the Coast Range, at the Golden Gate, allows a passage to the ocean. In the northern part of the State, where the streams from the Sierra run down to the Sacramento, this remarkable system of "buried river gravels" is found. In and before the Tertiary period of geology, these streams had worn valleys on the slopes of the Sierra, and made extensive deposits of gravel, by the erosion of the mountain-sides. Then came a period, or a succession, of volcanic disturbances and outflows, which made the great "lava beds" of northern California and Oregon. In many cases the lava flowed down and filled up the river-beds from side to side, covering the gravel deposits deeply, and often hardening and compacting them. When it had cooled sufficiently for normal conditions to be at all resumed, the drainage of the Sierra had to make its way by new lines. These were usually along the edges of the old valleys, on the top of the lava filling, at its junction with the sides of the former valleys. In the time that has since elapsed, these new valleys have been cut down deeper than the previous ones, at the expense of the intervening old divides; and the present condition is that the eastern affluents of the Sacramento are separated frequently by spurs running out from the Sierra, which consist at their top of the old gravels, more or less compacted, with a heavy protecting cap of lava or tufa. These old "sub-lava" gravels are those worked by the hydraulic process, or when consolidated into the so-called "cement" beds, by stamp-mills—all of them being gold-bearing and in some cases diamond-bearing. The surface gravels of the earlier prospecting and panning days are of course the work of the modern streams; they carry the placer gold, and occasionally a few diamonds.

There are some points in these occurrences that recall, at first sight, the diamond mines of Brazil and South Africa.

In Brazil the matrix is also a gravel, and is frequently cemented into a conglomerate ("cascalho") by oxide of iron. In Africa the diamond gravels contain associated minerals similar to those found in some of the California placers, notably in those of Butte County, where zircons, garnets, and rutile are met with. But these are not important relations, and afford no ground for assuming either a similar richness of yield or an identity of geological origin.

The earliest notice of a California diamond appeared very soon after the discovery of gold. It was published in the American Journal of Science, September, 1849 (II, vol. 8, p. 294), and relates how a clergyman from New England, the Rev. Mr. Lyman, had been shown an unmistakable diamond-crystal, of pale straw color, with convex faces, about the size of a small pea. He only saw it briefly, and the circumstances of its discovery, the exact locality, and what became of it are not known. In 1853 the first diamond was obtained from the Cherokee district, in Butte County, which has since been one of the principal localities. In 1854, Mr. Melville Attwood published an article in a newspaper, pointing out certain resemblances between the California deposits and the diamond gravels of Brazil, where he had long resided, and advising that search be made and care exercised, in view of the possible or probable occurrence of diamonds in the California gold-washings. From that time on, diamonds have been found at many points; though a far larger number of them have been lost or destroyed—either swept away by the violent current of the hydraulic mining, and buried in masses of débris, or crushed into fragments by stamp-mills. In 1867, Prof. Benjamin Silliman, Jr., exhibited several diamonds before the California Academy of Sciences, including one from the Cherokee district above noted, one from Fiddletown, near Volcano, in Amador County, another from El Dorado County, and still another from French Corral, Nevada County—most of them from the hard "cement" beds underlying the lava-flows. At the same meeting, Prof. J. D. Whitney enumerated some fifteen or twenty localities in the State that had yielded diamonds, adding that the largest stone he had seen was $7\frac{1}{4}$ carats; most of them being quite small.

The total number of California diamonds must now be quite considerable. Mr. H. W. Turner, of Washington, D. C., has recently summed up the facts in an article on "The Occurrence and Origin of Diamonds in California," published in the American Geologist, Vol. XXIII, March, 1899, pp. 182–191. He quotes from Prof. J. D. Whitney, and more recently from Mr. Henry G. Hanks, formerly State Geologist, who paid much attention to this subject, and gives a list of localities compiled from these two authorities. This list includes six counties, to which have since been added two others, Plumas and Tulare. These counties are (in geographical order) as follows: Del Norte, Trinity, Plumas, Butte, Nevada, El Dorado, Amador, and Tulare; making eight in all. Of these, Del Norte and Trinity stand apart, in the northwestern portion of the State, with streams flowing from the Coast Range into the Pacific; they have yielded only minute diamonds, in the sands of Smith River and Trinity River respectively. Tulare County, belonging in the San Joaquin section of the great California valley, is represented so far only by a single diamond, from Alpine Creek. The other five are all in the region drained by the Sacramento, as above described. El Dorado, Butte, and Amador have yielded the greatest number, and Nevada the largest stone—that of $7\frac{1}{4}$ carats, referred to by Professor Whitney—but only one or two others. Plumas County has two localities, Gopher Hill and Upper Spanish Creek, where a few small diamonds have been found in sands rich in heavy minerals, as in Trinity and Del Norte. The other three counties above named have furnished most of the diamonds of California. Amador contains the Volcano district, whence a number have come, among them one of the largest, a pale straw-colored crystal, weighing 255 milligrams ($1\frac{1}{4}$ carats). Butte County includes the famous Cherokee district, where sixty or more have been found; also Yankee Hill and Oroville, each of which has yielded several. El Dorado County has five or six localities near Placerville. Mr. Turner quotes Mr. George W. Kimble, of that place, as saying that most of the El Dorado specimens have come from "a point a little south of Smith's Flat and White Rock Cañon," in Neocene river-gravels; Webber Hill, in the same neighborhood, has also yielded several. From their geographical position, it would seem that similar diamond occurrences might be naturally expected in the counties of Yuba, Sierra, and Placer.

Comparing the South African occurrence of diamonds, in a serpentinous rock apparently derived from the alteration of a peridotite, Mr. Turner notes the fact that on the maps of the gold-belt, prepared by the U. S. Geological Survey, serpentine masses are indicated in the vicinity of all the above-named localities; he also cites Mr. Kimble, of Placerville, as stating that serpentine pebbles are frequent in the diamond-bearing gravels near that place, the rock itself outcropping four or five miles to the east. Mr. Turner suggests the examination of the gulches lying in these serpentine outcrops, as of interest and possible importance, with reference to the source whence the diamonds have actually come. This is as yet unknown; and though the African rock resembles a peridotite, or its decomposed and serpentinized product, yet this is by no means the only rock in which diamonds may occur. Those of Brazil, according to Prof. O. A. Derby's recent studies, are from rocks that are apparently metamorphic in origin, rather than igneous; and the whole problem awaits the results of further study.

Taking up the actual discoveries somewhat more in detail, and dating from the meeting of the California Academy of Sciences in 1867, before referred to, when Prof. Benjamin Silliman, Jr., and Prof. J. D. Whitney presented the subject fully, the principal facts are as follows, arranged (1) by counties, and in these (2) in order of time.

AMADOR COUNTY.—Of the several diamonds exhibited by Professor Silliman on the occasion mentioned,* one specimen, a little over 1 carat (3.6 grains) in weight, was from Indian Gulch, near Fiddletown; and four others from the same region were at that time known. These stones occurred in a compact volcanic ash or tufa, forming a gray "cement-gravel." At Volcano the rock is similar, and some sixty or seventy diamonds have been reported thus far. This is one of the places where the cement-rock is worked by stamping, and the tailings show pulverized diamonds. The crushed gravel pays well in gold; and it has not been thought desirable to change the present method and break up the rock in other ways more costly and troublesome, in order to save the diamonds that it may contain. In August, 1887, Mr. Hanks exhibited before the San Francisco Microscopical Society a beautiful stone of 1.57 carat (4.97 grains) weight, found at Volcano in 1882, belonging to J. Z. Davis, a member of the society, and now in the museum of the State Mining Bureau. It is a modified octahedron, about $\frac{3}{10}$ inch in diameter, transparent and nearly colorless, though slightly flawed. The curvature of the faces gives the crystal a sub-spherical form, but the edges of the pyramids are channels instead of planes. Closer examination shows that the channeled edges, the convex

---

* Proc. Cal. Acad. Sci., Vol. III, p. 354.

faces, and the solid angles are caused by an apparently secondary building up of the faces of a perfect octahedron; and for the same reason the girdle is not a perfect square, but has a somewhat circular form. These observations were well shown by enlarged drawings. The faces seem to be composed of thin plates overlying each other, each slightly smaller than the last. These plates are triangular, but the lines forming the triangles are curved, and the edges of the plates are beveled. Mr. Hanks remarked that under the microscope and by drawings exhibited it could be seen that each triangular plate was composed of three smaller triangles and that all the lines were slightly curved. The building up of plate upon plate caused the channeled edges and the somewhat globular form of this exquisite crystal. A close examination revealed tetrahedral impressions, as if the corners of minute cubes had been imprinted on the surface of the crystal while in

ILL. No. 7. Diamond, natural crystal, found at Volcano, Amador County.

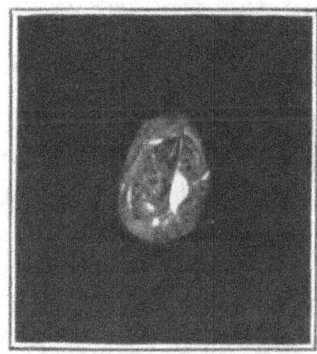

ILL. No. 8. Rough diamond, found in Spring Valley Hydraulic Mine, Cherokee Flat, Butte County.

a plastic state. These are the result of the law of crystallization, as was shown by the faint lines forming a lace-work of tiny triangles on the faces when the stone was placed in a proper light. Mr. Hanks concluded with the remark that it would be an act of vandalism to cut this beautiful crystal, which is doubly a gem, and he protested against its being destroyed by contact with the lapidary's wheel. Four small octahedral crystals taken from stamps at the Volcano locality were shown in the Tiffany collection of American precious stones at the Paris Exposition of 1889, and are now in the Tiffany-Morgan collection, at the American Museum of Natural History, New York City.

BUTTE COUNTY.—The Cherokee district, in this county, has been, from as early a date as 1853, one of the most prolific diamond localities in the State. Cherokee is near the North Fork of Feather River, and the geological relations of the diamonds and gold are essentially the same as those in Amador County, a hundred miles to the southeast, both districts lying among the western foothills of the Sierra, as previously described. Mr. Hanks called attention to included leaf impressions in the volcanic beds, as proving them to be tufas and not lavas. In number, the Cherokee diamonds obtained are about equal to those from Volcano. One was shown by Professor Silliman, on the occasion already mentioned, in 1867; and others were then known from that locality. William Brandreth obtained a crystal in the same year, which he afterwards had cut into a fine white stone of $1\frac{3}{16}$ carats. In 1873 several were obtained from the ground of the Spring Valley and Cherokee Mining Company, in cleaning up the sluices. One of these was described as large and straw-colored, while others were smaller, but very pure. Various stones, white, yellow, and pink, have from time to time been reported, and some have been cut and set. A fine crystal was presented to the State Museum by Mr. Williams, superintendent of the Spring Valley Mining Company. Two others, found at the same place in the summer of 1881, by Lucinda Voight, were shown by the present writer before the New York Academy of Sciences, January 12, 1886. Mr. H. S. Durden, of the California State Mining Bureau, reports that two small diamonds were obtained at Cherokee in 1892 and 1893, one of them weighing two carats.

Professor Silliman made the concentrations from the sluices of these Cherokee mines the subject of minute investigation, the results of which were published in two papers.* In the first he described his treatment of the material, both chemical and mechanical; and in the second he gives additional particulars, with results. He found here the following association of interesting minerals:—light-colored zircons, crystals of topaz, fragments of quartz, rutile, epidote, pyrite, and limonite, with some platinum, iridium, iridosmine, and gold, and a large quantity of black grains, which are proved by the magnet to consist about equally of chromite and titanite. At first he could find but little of the platinum and iridosmine, but this was due, as above stated, to the force of the hydraulic streams, which sweep away all small particles that do not amalgamate.

Mr. Hanks adds that platinum minerals have been found rather abundantly in Butte County. At St. Clair Flat, near Pentz, they were found in quantity in the early days of placer-mining. They are found also at the Corbier mine, near Magalia (Dogtown). As far back as 1861 a diamond was found one and a half miles northwest of Yankee Hill, in cleaning up a placer mine. It was taken from the sluice with the gold, and sold to N. H. Wells, to whom I am indebted for this information. He presented the gem to John Bidwell of Chico, who had it cut in Boston. It made a stone of $1\frac{1}{4}$ carats (4.75 grains). Mr. Bidwell

*See mineralogical notes on Utah, California, and Nevada, in the Eng. & Min. Jour. Vol. XVII, p. 148, March 11, 1873; and Am. J. Sci. (3), Vol. VI, p. 127, August, 1873.

gave it to his wife, who for years wore it in a ring. This was the only diamond found at this locality.

In 1895 Mr. Dwight Whiting reported the finding of five small diamonds near Oroville, on Feather River, and as many more about four miles from the head of the creek, suggesting a peridotite origin.

EL DORADO COUNTY.—Here a number of diamonds have been found at certain points. In 1867 Professor Silliman, at the meeting of the California Academy of Sciences, before mentioned, showed a crystal of 1½ carats (4.75 grains), of good color, though a little defective, from Forest Hill. It was found at great depth, in a tunnel run into the auriferous gravel. W. P. Carpenter, of Placerville, gave the following account of the locality in a letter to Mr. Hanks, in 1882: "In 1871, W. A. Goodyear, Assistant State Geologist, while examining the deposits of auriferous gravels in the ancient river-bed, about three miles east of Placerville, found several specimens of itacolumite, and expressed the opinion that diamonds should be found in the gravels. I assisted him in searching for them, and we found several in the hands of the miners. Mr. Goodyear bought one of them as a geological specimen. None of the parties who had them knew what they were, but kept them as curiosities. The gravel in the channel is capped with lava from 50 to 450 feet in depth. Of late years the gravel is worked by stamp gravel mills, and I know of instances where fragments of broken diamonds have been found in panning out the batteries."

He goes on to give the particulars of about fifteen diamonds obtained at different times in the neighborhood, some yellow and some white. One of these was a nearly spherical crystal, over one fourth of an inch in diameter, that was sold in San Francisco for $300, and another was sent to England to be cut. Subsequently in 1894, Mr. Carpenter announced that he had lately obtained two crystals, one weighing over 7 grains troy and the other 6, of rounded form and rough surface, each nearly one fourth of an inch in diameter and faintly tinted, the larger with a greenish shade and the smaller with pale yellowish. As many as forty or fifty small diamonds have been taken from the gravel at this place from time to time in the past, but since stamp-mills have been employed little is found but the crushed fragments encountered in "panning up" the amalgam taken from the batteries. Mr. Carpenter proposed to work his section of the channel by other means, to avoid the possible loss of diamonds of more value than the gold. The occurrence here is described as similar to that of most California diamonds— in the hard compacted gold-bearing gravel occupying ancient river channels now filled and overlain by igneous rocks.*

In the recently published article of Mr. H. W. Turner,† elsewhere

---

* Sixteenth Ann. Rept. U. S. Geol. Survey (1894), Part IV, p. 596.

† The Occurrence and Origin of Diamonds in California; by H. W. Turner. Amer. Geologist, Vol. XXIII, March, 1899, pp. 183, 184.

referred to, a letter from George W. Kimble, of Placerville, is quoted in regard to the diamonds of that vicinity. He states that two more had been found on the property of Thomas Ward & Co., on the south side of White Rock Cañon, where the stream cuts through beds of Neocene Tertiary, and nine others in the immediate vicinity, chiefly from the Unity mine, adjacent to that of the Ward Company; but no particulars are given as to their size or quality. Besides these, Mr. Kimble reports a great many fragments of crushed diamonds in the concentrates from the gravel mills. Later, 1898, he mentions having examined a diamond found within the limits of the town of Placerville, in Cedar Ravine, a tributary of Hangtown Creek. The name of Diamond has been given to a small railroad station a few miles south of Placerville.

NEVADA COUNTY.—Professor Silliman also showed to the California Academy of Sciences a very clear and symmetrical crystal from French Corral, Nevada County. It was thrown out of the cement-rock of deep gold washings, as usual, and weighed $1\frac{3}{5}$ carats (5.11 grains). The color was slightly yellowish; but this was perhaps due to its having been exposed to a red heat, as a test of its authenticity. Prof. Joseph D. Whitney, of Harvard College, stated, at the same meeting, that diamonds had been found in some fifteen or twenty localities in the State, and that the largest that he had seen was also from French Corral and weighed $7\frac{1}{4}$ carats.

In the northwestern counties of California, drained by the Trinity River, in the vicinity of Coos Bay, in Oregon, and on the banks of Smith River, Del Norte County, diamonds are occasionally found in the flumes and sluices. Some small ones are reported from Trinity County; and their mode of occurrence, similar to that of the diamonds of Cherokee district and of Oregon, is described in a letter to Dr. Charles F. Chandler, of the Columbia College School of Mines, from Prof. Frederick Wöhler, of Göttingen. He mentions having observed in the grains of native platinum from the sands of the Trinity River, Oregon, minute transparent zircons associated with laurite (sulphide of ruthenium and osmium), iridosmine, chromic iron, etc., and microscopic rounded crystals which he supposed were diamonds. In a subsequent communication, dated Göttingen, August 8, 1869, Professor Wöhler continues: "On examination under the microscope, the mineral powder which had been freed from platinum, gold, chromic iron (in part), silica, iron and tin, and from which the ruthenium, etc., had been removed by aqua regia, besides many grains of chromic iron and beautiful hyacinth crystals, colorless and transparent grains resembling quartz were observed, but besides these, grains resembling rounded diamond crystals were detected." He then describes in full his methods of testing these grains, and expresses his conviction that they were true diamonds.

CORUNDUM.

Ruby.   Sapphire.

H. = 9.   G. = 3.9–4.1.   Luster, adamantine to vitreous.   Colors include nearly all of the prismatic hues to colorless.   Dichroic.   Occasionally phosphorescent.

Oxide of aluminum = Al_2O_3.

The transparent corundums rank among the most valuable of gem stones, and include two standard varieties, the ruby and the sapphire.

Rubies are the red-colored corundums; also called *Oriental Ruby*. They vary in hue from a rose to a deep carmine, the choicest shade being called "pigeon's blood" red.

Sapphire in general includes all colors except red.   Accurately speaking, the name is limited to the blue colors, the choicest shades being known as royal blue, velvet blue, and cornflower blue.   *Oriental Emerald* is the green variety, ranging from a lively green to a bluish green.   *Oriental Amethyst* is the purple or *amethystine*.   These two are rare.   *Oriental Topaz* is yellow sapphire, rivaling the yellow diamond in brilliancy.   *Oriental Hyacinth* is honey red in tint.   *Adamantine Spar* includes the hair-brown varieties.   *Star Sapphire* has a stellated opalescence, as has the *Star Ruby*, which is also known as the *Asteria* or *Star-stone*.

The true ruby and sapphire are easily recognized by their hardness, as they can be scratched only by the diamond, but scratch all other stones.   They are also electrified by friction.

Corundum is associated with dolomite, gneiss, granite, mica, and chlorite slate.   The gems are usually obtained from placer workings.

Los Angeles County.—True sapphires have been found in the drift in San Francisquito Pass.

Plumas County is traversed at many points by large dikes of felsite and felsite porphyry.   This rock was first found by J. A. Edman and named plumasite by Lawson.   At a point near the western base of the serpentines, a large "pipe" of felsite outcrops, and in the soil near it are found fragments of feldspar containing corundum crystals, while between the serpentine and the felsite dike is a four-foot layer of the feldspar containing a few corundum crystals, small veins or strings of corundum ramifying into the feldspathic mass.   The largest crystal was two inches long by one inch wide, and of a bluish-gray color.   (S. M. B. 15357.)   No gem material has yet been found.

San Bernardino County.—At the east end of the Kingston range.

San Diego County.—Mr. W. H. Trenchard reports corundum, in opaque gray crystals, at a point some 26 miles east of San Diego, but in small amount and undeveloped.

TOPAZ.

H.= 8. G.= 3.4–3.6. Brittle. Has very perfect cleavage (basal) transverse to the elongation of the crystals. Luster vitreous. Color various shades, yellow to brown, and pale green or blue, but often colorless or with a faint tinge of bluish green. Silica, 33.3; alumina, 56.5; fluorine, 17.6.

Usually occurs in rhombic prisms, the crystals more or less pointed. The color of some yellow topazes fades on exposure to sunlight, while others—especially those from Brazil—change their yellow color to pink on heating.

SAN DIEGO COUNTY.—Beautiful topazes have lately been found near Ramona. Some of the crystals are colorless, others are bluish or greenish. Many are quite large and are covered with numerous small etch-figures. Some crystals found were over two inches long and one inch wide.

ILL. No. 9. TOPAZ CRYSTAL (NATURAL SIZE), RAMONA, SAN DIEGO COUNTY.

The mines that have yielded this very interesting addition to the gem-minerals of California are chiefly those known as the Surprise and the Little Three, adjacent to each other, about $4\frac{1}{2}$ miles northeast of Ramona. The topaz occurs in a pegmatite ledge, of the kind elsewhere described as characteristic of this region, but not in all parts of it, being met with only in certain portions of the vein or dike. At the Little Three mine they occur in pockets in albite and orthoclase with quartz crystals; they are attached to the feldspar, and surrounded with a red soil that fills the pockets. Associated with them are dark-green tourmaline crystals, sometimes very large. The topazes are white, light yellow, sea-green, and sky-blue, and some crystals are over a pound in weight.

At the Surprise mine, in the same way, topaz is found only in one part of the workings, and its occurrence is described as quite similar.

In the central part of the pegmatite vein, between the crystallized upper portion and the more compact portion below, lies a zone of small pockets in feldspar (in this case orthoclase), somewhat decomposed. In these are found the topaz crystals, in a sandy filling of granular ferruginous quartz. Those near the surface were colorless or white, but at a depth of 6 feet they were sky-blue and aquamarine-blue. Some 50 pounds of them have been taken from a cut 20 feet long and 8 feet deep.

Illustrations Nos. 9 and 10 show crystals of natural size, some of them surrounded with the finely crystallized albite, from Ramona.

ILL. No. 10. TOPAZ CRYSTALS (NATURAL SIZE) ON CRYSTALLIZED ALBITE, RAMONA, SAN DIEGO COUNTY.

SPINEL.

Spinel is a compound of alumina and magnesia, often with some iron, or other metallic oxides in small quantities. It has a hardness of 8, and when transparent makes a valuable gem-stone, usually of rich shades of red, and is then called spinel ruby, or ruby spinel. It has not been found much in California, but some crystals of good quality, yielding gems weighing up to two carats each, were obtained near San Luis Obispo, by Mr. James W. Beath, of Philadelphia, Pa. In the collection of the State Mining Bureau there are also crystals of wine-colored ruby spinel up to 3 millimeters in diameter, from Humboldt

County. Crystals of blue spinel, of about the same size, clear and of good color, have lately been obtained sparingly from the Mack mine near Rincon, San Diego County.

BERYL.

Emerald, Aquamarine, Goshenite, etc.

H. = 7.5–8, scratching quartz, but scratched by spinel or topaz. G. = 2.7. Brittle. Luster vitreous. Colors, emerald green to pale green, pale blue, pale yellow, honey, wine, and citron yellow, white to a pale rose-red. Silica, 67.0; alumina, 19.0; glucina, 14.0. The crystals are usually hexagonal prisms, occasionally very large, but those of fine quality or transparency are apt to be small. They are electrified by friction.

The emerald and aquamarine are mineralogically included in the species beryl; the differences being mainly in color, as follows: *Emerald* includes the rich green kinds only, and is a highly prized gem when free from flaws; *Aquamarine* includes the beryls showing clear shades of sky-blue and sea-green; *Goshenite*, white or colorless; *Davidsonite*, greenish yellow; *Aëroides*, pale sky-blue; *Hyacinthozontes*, clear sapphire blue; *Améthiste Basaltine*, pale violet or reddish; *Chrysolithus*, pale yellowish green; *Golden Beryl*, clear bright yellow; *Chrysoberyllus*, greenish yellow to wine-yellow. Occasionally, beryl occurs of a delicate pink color.

A number of localities for beryl are known in California, some yielding gem material. The pink or rose beryl, hitherto one of the rarest varieties of this species, has within a year or two past been found at several points in the remarkable mineral region of San Diego and Riverside counties, sometimes of transparent gem quality, and in a few cases of large size, as noted further on. With it, and also separately, are found beryls of other kinds—green, yellow, and colorless. At the Mack mine at Rincon, San Diego County, near Oak Grove, are also reported peculiar beryls of a deep opaque blue.

It is interesting to note that this pink or rose beryl occurs with the pink and lilac spodumene at several California mines, while the beryls are rich green when associated with emerald-green spodumene (hiddenite) at the remarkable locality at Stony Point, Alexander County, North Carolina, and even the muscovite associated with them has sometimes a green color. That is, the beryls occur there in the form of emeralds, the spodumene in the form of hiddenite, and the muscovite mica in a beautiful green tint, evidently all due to some chromium coloring. This latter (muscovite) is unusually interesting when embedded in transparent quartz, and one can see through the sides of the prisms.

In contrast to this, at several localities in southern California the beryls are pink in color, the rubellites deep pink, the spodumene (variety kunzite) lilac, and the lepidolite also of pink and lilac tints.

The most important beryl occurrences in California are the following:

RIVERSIDE COUNTY.—Associated with the tourmalines from Coahuila have recently been found yellow beryls, closely resembling those from Sarapulka in the Ural Mountains, also pale green, pink, and colorless. Some of the yellow crystals are finely formed and others show remarkable instances of etched faces, while others are almost as delicate as a darning-needle.

SAN DIEGO COUNTY.—The notable locality of colored tourmalines and other lithia minerals, the Himalaya mine, at Mesa Grande, has yielded a most unusual specimen of pink beryl—a transparent rose-colored mass, measuring 65 by 50 millimeters. It is evidently an etched fragment of a very large crystal, showing planes with markings and erosions all over its surface. Its color varies by transmitted light from a delicate rose to a deep rich pink. This beautiful specimen is now in the Tiffany-Morgan collection in the American Museum of Natural History in New York City. Another somewhat similar rose beryl has lately been obtained from Mount Palomar near Oak Grove. This crystal measures 11 cm. by 7½ cm. by 6½ cm., and weighs almost 2 pounds. It is perfectly transparent and of a beautiful pale-rose color. It is now in the United States National Museum.

From Pala, in the same county, occurring with tourmalines and kunzites, was obtained a large detached rose beryl measuring 10 cm. by 5 cm. This crystal was of a pale pink color, and transparent.

Other localities of pink beryl, lately reported, are as follows:—

The Esmeralda mine, at Mesa Grande, together with golden beryl and aquamarine; both pink and green at the Crystal gem mine, near Jacumba; and particularly in some of the mines near Ramona. Here, the Surprise mine reports two pounds of large and fine crystals; and the A B C mine several pounds, from which many choice gems have been cut in San Diego, one of them a flawless stone of 30 carats, rose-leaf pink in color. Beryls of more ordinary tints, of various shades of green, are reported from a number of the mines of San Diego and Riverside counties, sometimes of choice quality, especially the Fano mine in the latter county, the Hercules and Lookout mines at Ramona, and the Mack mine at Mount Palomar, in the former.

From near Ramona have also come some very curiously etched crystals; some of three inches long and an inch across, colorless and transparent as the finest rock-crystal, and covered all over the prismatic and basal planes with the most complicated etching; these are hollow within, and made up of interlocking plates, as it were, exceedingly clear and brilliant. Pink beryls have also been found here.

The mine owned by J. M. Mack, of San Diego, is situated in Sec. 25, T. 10 S., R. 1 W., S. B. M., on the foothills of Palomar Mountain, 9 miles southeast of Pala, on a direct line between that place and Mesa Grande. Beryls are found here of various colors—yellow, green, etc., to deep blue—with columbite, and much crystallized quartz and feldspar, also black tourmalines, but none of the colored varieties so marked at Mesa Grande and Pala. Some of the beryl crystals are very small and slender, but extremely perfect and brilliant; others were singularly eroded, as though acted upon by some solvent; others were like certain beryls from Haddam Neck, Conn., with part of the crystal transparent and the rest cloudy or milky, curiously resembling a test-tube in which a white precipitate is subsiding from a clear green liquid, the line of demarcation being quite sharp. The beryls here are found in pockets in a pegmatite rock, like all the gem-minerals of this region, usually embedded in a red clay. Many fine gems have been cut from the clear crystals, and the clouded ones have been cut *en cabochon*, and have a pleasing cat's-eye effect.

GARNET.

Grossularite, Pyrope, Almandite, Spessartite, Andradite, Ouvarovite, etc.

The species garnet includes several varieties which are distinguished by differences in composition. The group in general has :—

H.=6.5. G.=3.15–4.3. The majority of specimens will scratch quartz slightly. They are complex silicates of alumina, lime, magnesia, chrome, iron, manganese, or titanium, grouped as follows :

1. Aluminum garnets : Grossularite = Lime-Aluminum garnet; Pyrope = Magnesium-Aluminum garnet; Almandite = Iron-Aluminum garnet; Spessartite = Manganese-Aluminum garnet.

2. Iron garnet : Andradite = Lime-Iron garnet.

3. Chromium garnet = Lime-Chrome garnet.

1. *Grossularite* has a hardness of 7, and G.= 3.55–3.66. Color white, pale green, amber, honey, wine and brownish yellow, cinnamon-brown, and pale rose-red. The varieties of grossularite are essonite, or hyacinth, and cinnamon stone. The essonite is the only true hyacinth of the jeweler, and has been confused by some with zircon. In the trade the name grossularite is confined to the pale-green or yellow stones; and cinnamon stone to the cinnamon-brown color. *Romanzovite* is brown; *Wiluite* is yellowish green to greenish white; *Topazolite* is topaz to citron yellow; and *Succinite* is an amber-colored kind of grossularite. A pink variety called *Rosolite* occurs in Mexico.

Pyrope (meaning "fire-like") is the principal magnesian garnet, a deep blood-red to nearly black stone, prized as a gem. H.= 7.5. G.= 3.7–3.8. It is known as the precious garnet.

Almandite. H.= 7.5. G.= 3.9–4.2. Iron-aluminum garnets prized as gems are also called precious garnet, like pyrope. Color cherry-red, blood-red to deep reds. Almandite is often called *Carbuncle* when of a deep clear red, scarlet or crimson. The true carbuncle is a variety of sapphire.

The variety *Rhodolite* has a color between violet purple and brownish red, and belongs between almandite and pyrope.

Spessartite is a manganese-aluminum garnet. H.= 7. G.= 4.0. The color varies from a reddish brown, or dark hyacinth red, and even violet, to orange red. It often affords fine gems.

2. *Andradite*, or lime-iron garnet. H.= 7.5. G.= 3.8–3.9. The group includes varieties that differ widely in composition and color. The trade name Andradite is limited to the yellow or orange-brown variety. *Demantoid* or *Uralian Emerald* is a grass-green, emerald-green, or brownish-green stone having a brilliant luster, and showing good fire when cut. *Colophonite* is a brownish-black variety, characterized by a resinous luster. *Melanite* is black to yellow-brown.

3. *Ouvarovite*, or *Uvarovite*, lime-chrome garnet. H.= 7.5. G.= 3.41–3.52. The color is a fine emerald green.

Trautwinite is an impure uvarovite from Monterey County, described by E. Goldsmith in Proc. Acad. Nat. Sci. Phila., 1865, pp. 9, 348, 365.

Schorlomite has a composition analogous to garnet. H.= 7–7.5. G.= 3.81–3.88. Color black, sometimes mixed with blue.

In the trade but little attention is paid to mineral differences, but the garnets are often classified by color, the light-colored clear ones being called *hyacinth;* the yellowish, *jacinta;* a yellowish-red, *guarnaccine* or *vermeille;* the red with a tinge of violet, *rubino de rocca* or *grenat siriam;* and the deep clear red, *carbuncle*, especially when cut *en cabochon.*

The almandite is common in granite, gneiss, and mica schist; grossularite is frequent in limestones and crystalline schists; pyrope in peridotites, serpentines, and basalts; spessartite in granite rocks, quartzite, schists, and rhyolites; iron garnets in eruptive rocks; demantoid in serpentine; chrome garnets with chromite in serpentines and in granular limestones.

CALAVERAS COUNTY.—Almandite, from Bald Point, Mokelumne River. S. M. B. 11857.

EL DORADO COUNTY.—Grossularite in copper ore, Rodgers mine, in the eastern part of the county; associated with specular iron, calcite, and iron and copper pyrites. Garnet rock is found in blocks several feet thick near Pilot Hill. S. M. B. 13937.

FRESNO COUNTY.—In calcite at San Ramon, S. M. B. 9336; in feld-spathic rock at Grub Gulch, S. M. B. 7037; with epidote on quartz, at Fresno Flats, S. M. B. 7317.

INYO COUNTY.—Garnets are found in the Coosa district in large, semi-crystalline masses, of a light yellow color. Grossularite with datolite occurs at San Carlos, S. M. B. 2190. Cinnamon-stone is also found at San Carlos.

KERN COUNTY.—Garnet sands are abundant at the Soapstone Mountain and in the Mojave Desert. S. M. B. 2882.

MARIN COUNTY.—Garnets in mica schists, Reed's ranch. S. M. B. 6562 and 12833.

MARIPOSA COUNTY.—Almandite at Mount Hoffman. S. M. B. 12007.

MONTEREY COUNTY.—Trautwinite, locality not given. Pyrope in granite, from Nacimiento River. S. M. B. 13726.

PLUMAS COUNTY.—The late Dr. Isaac Lea, of Philadelphia, whose great collection of precious stones is now in the U. S. National Museum at Washington, had some transparent crystals of a dark oily green grossularite from 1 to 5 millimeters long, that were found at the Good Hope mine.

RIVERSIDE COUNTY.—Essonite garnet in handsome crystals has been found at some of the tourmaline localities near Coahuila, as noted below.

SHASTA COUNTY.—Ouvarovite, from Shotgun Creek. S. M. B. 11729.

SONOMA COUNTY.—Grossularite in copper ore near Petaluma.

SAN BERNARDINO COUNTY.—Garnets are common in the placer sands of the desert. Grossularite, S. M. B. 6614.

SAN DIEGO COUNTY.—Almandite occurs in mica schist, at San Margarita ranch. S. M. B. 12233.

Essonite and succinite appear at a number of localities in deposits spread over a considerable territory from 9 to 10 miles northeast of Jacumba Hot Springs, usually associated with granite and granular limestone. At three of the places some gem material has been found. Associated with it is a little vesuvianite and crystallized quartz. Eleven localities in this region are noted by Mr. W. H. Trenchard, of San Diego. Essonite has also been found near San Vicente, El Cajon Mountains, but the crystals were full of imperfections. The finest essonite crystals are obtained at Ramona, implanted on feldspar, and associated with green tourmaline, white topaz, and beryl, occasionally in perfect dodeca-hedrons and trapezohedrons, of rich honey-yellow to orange-red color, and very brilliant. They have also been discovered at Warner's Ranch, Mesa Grande, Santa Ysabel, Gravilla, and Julian, San Diego County; Deer Park, Placer County; Laguna Mountains and Jacumba, and also at several places below the Mexican line. As some of the crystals were of exceptional brilliancy, it is possible that on further development many fine gems will be obtained. This essonite garnet has been confounded with spessartite, and frequently reported as such, but it is really the former species in most, if not all, cases in this region. Among the mines in the Ramona district yielding specimens of very fine quality are the Hercules, Lookout, Surprise, and Prospect, from some of which beautiful gems have been cut, ranging up to 6 or 8 carats; also excellent quality near Jacumba, and of less size and beauty at many places. Deep red garnets, 6 to 10 millimeters in diameter, also occur in this region. Larger ones, up to as much as 30 millimeters, occur near Coahuila, Riverside County, in trapezohedral crystals of remarkable beauty.

SANTA CLARA COUNTY.—A cinnamon-stone from this county, analyzed by J. L. Smith, gave silica, 42.01; alumina, 17.76; ferric oxide, 5.06; manganous oxide, 0.20; lime, 35.01; magnesia, 0.13. G.=3.59.

TRINITY COUNTY.—Richly colored ouvarovite was discovered in 1899 by Mr. George L. Carr and others, at Carrville. It occurs in small dodecahedral crystals from 1 to 3 millimeters in diameter, of the richest deep green, coating seams or cavities in chromic iron. These were at first thought to be emeralds, until analysis proved their real character.

TULARE COUNTY.—Several varieties of garnets occur in this county, at various points. These have been principally reported by Mr. Max Braverman, who searched and explored for minerals in this region with indefatigable activity, and whose collection, generously presented by him in 1901 to the Golden Gate Museum, remains as a worthy record of his many years of labor and enthusiasm. He has reported essonite at Three Rivers, pyrope on Rattlesnake Creek, and topazolite from near the chrysoprase locality, 12 miles northeast of Visalia. In 1900 many fine groups of crystals were taken out at this locality. Almandite was reported as abundant between North and Middle Tule rivers, by Mr. L. B. Hawkins. Two specimens of topazolite, with malachite and azurite, were exhibited by Tiffany & Co. at the Paris Exposition of 1889, that came from this county. A curious white mineral, associated with the massive green vesuvianite (californite), was obtained by Mr. Braverman in 1902, from a point in this county near the Fresno county line, not far from Selma, a mile and a half from Hawkins schoolhouse. This proved, on analysis by Mr. George Steiger, of the U. S. Geological

Survey, to be a massive variety of grossularite garnet—a peculiar and unusual form. The analysis is as follows:*

SiO_2	38.59
Al_2O_3	22.24
Fe_2O_3	0.45
FeO	0.36
MgO	0.64
CaO	35.97
MnO	0.10
H_2O (below 100° C.)	0.31
H_2O (above 100° C.)	0.80
CO_2	0.39
F	0.17
	100.02

VENTURA COUNTY.—Garnet sands are abundant in the Piru district. S. M. B. 2365.

Garnets are abundant in all the counties where the gravels are worked for gold, and are generally called "rubies" by the miners.

TOURMALINE.

Rubellite. Indicolite. Achroite. Aphrizite.

H. = 7–7.5. G. = 3.0–3.2. Brittle. Luster vitreous. Color black, brown, blue, green, red, colorless. Some specimens are red internally and green externally; others red at one end and green, blue, or black at the other.

The red or pink transparent varieties are called *Rubellite;* if violet-red, *Siberite; Indicolite* is blue or bluish black; *Brazilian Sapphire* is Berlin blue; *Brazilian Emerald, Chrysolite of Brazil*, green and transparent; *Peridot of Ceylon* is honey-yellow. *Achroite* is the name given to colorless tourmalines. *Aphrizite* is black, with a resinous fracture. *Dravite* is brown, greenish black.

A complex silicate of alumina, boron, magnesia, iron, and alkalies (soda, potash, lithia), with small amounts of water and fluorine.

Among the most interesting and beautiful of gem-minerals are the highly colored varieties of tourmaline. It is only recently that they have come to be much known or used in jewelry, though pink (shan) tourmaline has long been greatly and almost superstitiously prized in China. The ordinary tourmalines are black or brown, but some varieties are pink, red, green, and dark blue, and these when transparent make elegant gems. Rarely, they are quite colorless (achroite). Tourmalines are remarkable also for certain optical properties which render them incapable of being successfully imitated, and for the fact that the same crystal will often show two or more richly contrasting colors in different parts.

* U. S. Census Rept. 1900, Precious Stones, by G. F. Kunz, p. 1050.

California has lately been found to possess the most remarkable mines of these gem-tourmalines in the world. Heretofore they have come chiefly from Brazil, and also from Oxford County, Maine, and Haddam Neck, Conn. Now, however, there are several points in San Diego and Riverside counties that are yielding splendid material.

The first recognition of these minerals in the State apparently goes back as far as 1872, when Mr. Henry Hamilton, in June of that year, obtained some very fine and handsome colored tourmalines on the southeast slope of Thomas Mountain, in Riverside County.

The first discovery in San Diego County is thought to have been made about twenty-five years ago, when some Indian children, at play in a camp near what is now Mesa Grande postoffice, picked up an oddly shaped stone, six-sided like a quartz crystal, about three inches long and a little thicker than a common lead-pencil. On cleaning it off and rubbing it with a bit of hide, it was seen to be of a beautiful blue color, bright and partially clear, almost like a sapphire. The natives had no idea of its nature, but were attracted by its beauty and singularity. Subsequently, other highly colored stones of like character—some blue, others green, others red—were picked up in the same vicinity by Indians and cowboys, but no one realized that they had any actual value.

In Pala, San Diego County, Mr. William Irelan, Jr., State Mineralogist, reported that fine transparent crystals of rubellite (red tourmaline), though not of gem quality, had been found.

The first important development at Pala was announced by Mr. C. R. Orcutt, in 1890.* Here a ledge of lepidolite (lithia mica) containing rubellite was traced for quite a distance. The rubellite crystals are clustered in radiating groups in the fine compact mica; they are not large and not clear, and hence are not suitable for cutting; but their color is a rich rose-red, and they make elegant specimens, on the background of lilac lepidolite. These have gone into collections and museums all over the world, and the material has been mined by tons, partly for specimens and partly for extracting lithia compounds from the lepidolite. More recently, amblygonite (alumina-lithia phosphate) has been found at this mine in large quantities, and this is now the greatest lithia mine in the world.

The rubellite crystals found here are entirely embedded in lepidolite, and until recently it was found impossible to remove them to show their complete form. They were, however, often polished with the lepidolite— the rubellite appearing as pink radiations in a darker gangue of lilac-colored lepidolite. Recently, however, the crystals of rubellite have been worked out, as it were—made to stand out by removing the lepidolite matrix by means of brushes and cleaning-tools—forming most beautiful groups of crystals.

---

*Report on the minerals of the Colorado Desert; 10th Ann. Rept. State Mineralogist of California, 1890. Min. Res. U. S. (Rept. U. S. Geol. Survey), 1893, pp. 17, 18 (reprint).

In regard to the early history of this locality, Mr. F. M. Sickler, who grew up in the vicinity and has explored for mines and minerals thereabout a great deal, relates the following curious and somewhat romantic circumstances, in an article in the Kansas City "Jeweler and Optician," of May, 1904. He states that the Pala lepidolite deposit had very long been known to the Indians, but that it was first brought to the notice of the whites by an Indian deer-hunter named Vensuelada. He found the spot while hunting, and broke off pieces showing the beautiful pink rubellite in its matrix of pearl-colored lepidolite, and brought them to Pala. Henry Magee, an old miner and prospector, took the rubellite

ILL. No. 11. TOWN OF PALA, SAN DIEGO COUNTY—VIEW LOOKING EAST, SHOWING INDIAN HOUSES.

crystals for cinnabar, and located the property as a quicksilver mine. Failing to get any mercury from it, he nevertheless believed that the peculiar mineral must have some value, and sent samples to various chemists, but no one recognized it as a lithia compound of any importance. Weary of his poor success, Magee gave it up and failed to do the annual assessment work on the claim. Later, one Tomas Alvarado relocated the property as a marble quarry! Magee claimed that some interest in the mine was rightfully due to himself, but Alvarado refused to give him any. Upon this, Magee pointed eastward to the ridge now called Heriart Mountain, and said, "If this stuff is of any value, I know where there are thousands of tons of it over there." Magee died, however, and his secret died with him; but certain it is that several mines, with lepidolite and tourmaline, have lately been located on that very ridge.

In 1893, near the crest of the San Jacinto range, in Riverside County, loose or "float" crystals of tourmaline were observed, chiefly black, but some finely colored—red, rose, green, blue, etc.* In some cases, the green crystals were found to have red centers—a type long known from Brazil. Some large crystals were obtained and a number of gems were cut from them. These indications were promptly followed up, and several mining claims were located and worked.

ILL. No. 12. BELFRY OF OLD SPANISH CHURCH AT PALA, SAN DIEGO COUNTY.

One of these, opened near the summit of the range by three prospectors, Messrs. Dwight Whiting, F. M. Speer, and F. H. Jackson, was called by them the San Jacinto gem mine. It was reported that more than a bushel of red and green crystals was found during the first season's operation, one of which measured eight inches in length and several inches in diameter. This was purchased by Harvard University, with other crystals several inches long and two inches in diameter. One of this size had a dark green basal termination and showed a red center on the fracture at the other end of the crystal. Other very fine ones are in the American Museum of Natural History, at New York.

*G. F. Kunz, U. S. Geol. Survey, Min. Res. U. S., 1892, p. 12 (reprint).

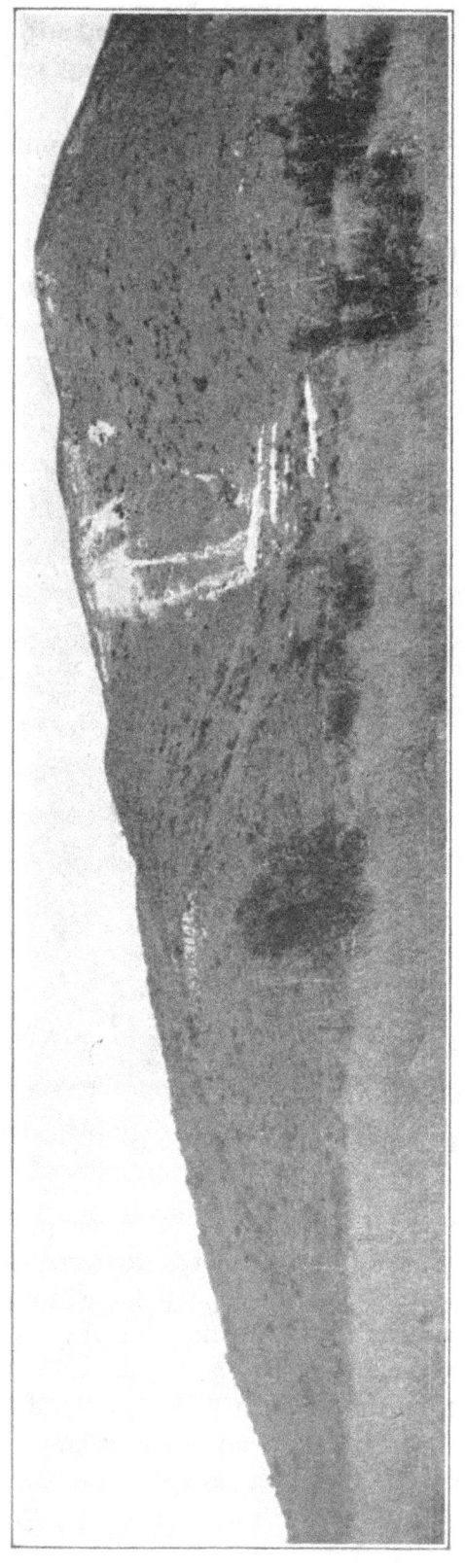

ILL. No. 13. PALA MOUNTAIN—VIEW OF THE LITHIA MINE, SHOWING WORKINGS AND DUMP. LEPIDOLITE WITH RUBELLITE.

Elegant specimens were made by cutting and polishing sections across the prism, in some of the large crystals of this type, showing the rich green exterior, then a narrow zone of white, and within that the red central portion— a beautiful contrast of colors, recalling a slice from a watermelon. Some of these were as much as three inches in diameter. A few years later, remarkably fine crystals of colorless tourmaline (achroite) were reported from this locality, by Mr. Dwight Whiting.

Soon there were several mines in operation in the San Jacinto district, and these gave quite a valuable output for many years. At present the one that is most prominent is that known as the Fano (formerly the Simmons) mine, discovered in 1902 by Mr. Bert Simmons, but now owned by Mr. E. A. Fano, of San Diego. This is located on the north side of Coahuila Mountain, at an altitude of some 4500 feet, about a mile south of Bautista Creek, and four miles west of Ramona Indian reservation. About the same distance east of the reservation, and a little south, on Thomas Mountain, at 5000 feet elevation; is the site of the original discovery of colored tourmalines in this county, made by Mr. Hamilton in 1872. This has been known as the Columbia gem mine, and has yielded very fine material; but it has not been worked much of late years, owing to litigation, other parties claiming it under the name of the April Fool mine.

The Fano mine, besides colored tourmalines, especially rich shades of blue and green, yields some beryl and a little kunzite, with lepidolite and amblygonite. Its structure is typical of the gem deposits of southern California—a ledge consisting of a vein (or dike) of pegmatite, about five feet thick, with a northwest and southeast course, and a dip of 17° to the southwest. The inclosing rock is called a blue granite, but is probably the diorite (or gabbro) rock.

In 1895, a rubellite crystal was reported as found on the Dameron place, in San Diego County, about 25 miles southeast of Pala, and a mile northwest of the Indian rancheria, at Mesa Grande. The pink tourmaline was noted as associated with the black variety in the rock of certain coarse crystalline granitic dikes (pegmatite) of that district. As these dikes are frequent, search was made among them; and in 1898, the great tourmaline locality of Mesa Grande was located.* The occurrence has many resemblances to that at Pala, and also marked differences.

The tourmalines are in large and distinct crystals, often transparent, sometimes in lepidolite, sometimes in quartz, and sometimes in feldspar—more as in Maine and Connecticut, and in the Riverside County mines. As at Pala, the red variety predominates, but there are many crystals of other tints—blue, green, etc., and perfectly colorless—and many that are parti-colored; others are red externally and green internally, like the Brazilian, but reversed.

The vein or dike in which the tourmalines occur here is at an altitude of 5000 feet, on the same belt as that which contains the similar minerals at Pala, on Smith's Mountain. The spot seems to have been known to the Indians, probably from the incident before noted; and they have even done a little crude blasting to break up the rock and procure the colored crystals from the cavities in which they chiefly occur.

The Mesa Grande locality is remarkable for the great size and perfection of the crystals, many of them being almost faultless, and the doubly-terminated ones being the rule rather than the exception. The Ernest Schernikow collection from this mine is the finest known, and ranks with those from any locality in the world. It has recently (1904–05) been on exhibition at the rooms of the State Mining Bureau in San Francisco. A very fine set of specimens was shown at the Pan-American Exposition at Buffalo in 1901, in the Tiffany exhibit of American gems, since purchased and presented to the Musée d'Histoire Naturelle at Paris. The two-color crystals, part green and part red, are remarkably strong at the point of contact, so that many have

* Min. Res. U. S., 1900, pp. 33, 34 (reprint).

been cut showing one-half of the gem green, either pale or dark, and the other a handsome pink or red. Some of the crystals have circular hollows or threadlike inclusions, so that when they are cut across these layers they form admirable cat's-eyes. Quite a number of remarkably beautiful cat's-eyes weighing from 25 to 30 carats have been found, varying from almost colorless to pale pink, rose, red, pale green, yellow-green and dark green. For this peculiar type of tourmalines, Mesa Grande is preëminent. One very large gem, weighing over 50 carats, showed the beautiful bi-coloration of pink and green. This locality has been worked more thoroughly and has been more productive than any other in the United States. The mineralogical specimens alone must have a value of some $30,000; and up to 1905, gems to the value

ILL. No. 14. RED TOURMALINE CRYSTALS ON QUARTZ CRYSTAL,
MESA GRANDE.

of $200,000 are said to have been taken out. A fine series of these crystals is in the Morgan collection in the American Museum of Natural History, New York.

During the past year or two several other remarkable localities have been discovered. One of these, the Pala Chief mine, is situated a mile and a half northeast of the town of Pala, and within a mile of the celebrated lepidolite and rubellite workings at that place. But the tourmalines at the new opening are more like those of Mesa Grande, and even larger. Some crystals were as much as a foot long and three inches across, of rich pink rubellite with an exterior coating of the dark blue variety, indicolite, separated by a pale intervening zone. Other pink crystals have a blue cap or termination, of a deep shade, inclining toward purple. One very remarkably large crystal is like a hollow cylinder, apparently composed of a group of prisms surrounding an open central space at the axis of the cluster; this is entirely of a rather dull blue, verging toward reddish in the interior. This locality, however, has not yet been worked for its tourmalines, owing to its yielding the remarkable new gem-stone, kunzite—transparent lilac spodumene, which has attracted so much attention of late.

The principal mine at this locality, known as the Pala Chief, was located in May, 1903, by Mr. Frank A. Salmons, John Giddens, and

ILL. No. 15. PALA CHIEF MINE, PALA—SOUTHWESTERN END OF WORKINGS, LOOKING EAST. KUNZITE AND GEM-TOURMALINE.

two Basque French prospectors, Bernardo Heriart and Pedro Peiletch. It presents the usual type of the mines of this region—a large vein or dike of pegmatite, between upper and lower walls of gray, somewhat decomposed, diorite (or gabbro). The vein has the characteristic division, elsewhere described, into an upper portion, more or less coarsely crystallized, consisting largely of the two feldspars, albite and orthoclase, with some quartz, and a lower portion of fine compact feldspathic granite, without mica, and lined or banded in layers with small essonite garnets—the so-called "line-rock." Between these is a zone of pockets, with much lepidolite, and the pockets filled with a talcose or clay-like material, of white and reddish tints, in which are found the tourmaline and kunzite crystals. These have been already described.

On another ridge eastward of this one, but separated only by a narrow valley, the two Basque Frenchmen above named, in conjunction with Mr. Fred M. Sickler and his father, M. M. Sickler, have discovered and located several claims showing kunzite and more or less of gem-tourmaline. The name of Heriart Mountain has been given to this ridge, which is apparently a foothill or spur of Agua Tibia Mountain.

Several tourmaline mines have also been located north and east of Coahuila, Riverside County, by Mr. Bert Simmons, of Oak Grove, from which some large gems have been obtained. Mr. Simmons has also developed several mines southwest of Oak Grove, on Aguanga Mountain, which have produced a number of gem-tourmalines. From one of the mines here, owned by Mr. K. C. Naylor of San Diego, several fine yellow stones have been taken. These mines have also shown a number of other minerals, such as spodumene (kunzite), columbite-tantalite, pink beryl, etc. The first announcement of this region was made by the writer.*

This entire lithia region, the geological character of the deposits as well as the descriptive mineralogical part, is now being monographed by Dr. Waldemar T. Schaller, of the U. S. Geological Survey,† to be published in 1905.

Another discovery of a locality for colored tourmalines and kunzite is recently reported from a new district in San Diego County, about 10 miles south and somewhat west from Pala, in Section 26, T. 10 S., R. 3 E. The locality is in Moosa Cañon, near Moosa Falls, and the discoverers are Messrs. Thomas Freeman and Joe Meyers, of Oceanside. The surface indications, and specimens brought in, suggest that the discovery may prove as rich as the other and older mines. Quartz crystals of large size are also found there, and smoky topaz is said to be abundant; but this is more probably smoky quartz, as the topaz has not been found associated heretofore with kunzite or highly colored tourmaline.

These gem-tourmalines all contain some lithia, and are found in association with other lithia minerals, such as lepidolite (lithia mica), amblygonite (lithia phosphate), and the alumina-lithia silicate, spodumene (kunzite). The particular associations, however, vary at different localities. Thus, in the Pala district there are three adjacent ridges; the western one, known as Pala Mountain, contains the great lepidolite and amblygonite mine, now worked for some years, which has furnished the radiating groups of pink rubellite above

---

*Report Dept. Min. Statistics, U. S. Geol. Survey, 1901, p. 31 (reprint). Science. January 28, 1904 (Vol. XIX).

†Science, February 12, 1904, p. 266 (Vol. XIX).

described, elegant as specimens, but not transparent; on the same ridge is the Stewart mine, which yields larger crystals of rubellite and some of other colors. On the middle ridge, Pala Chief Mountain, is the recently opened mine of that name, already mentioned, where the very large colored tourmalines occur, with kunzite, but little or no amblygonite. Other openings on this ridge are the Tourmaline Queen and Tourmaline King, of which the former especially shows crystals of rich and varied coloring. On the eastern ridge, Heriart Mountain, are several openings at which kunzite is found, and frequent association of

ILL. No. 16. PALA MOUNTAIN, SAN DIEGO COUNTY, LEPIDOLITE MINE. NEAR
VIEW OF THE DUMP.

gem-tourmaline and lepidolite. The Mesa Grande locality has already been described; the great mine is the Himalaya, but the Esmeralda and one or two others have rich and beautiful gem-tourmalines, but none have kunzite. The Ramona district, that yields garnet and topaz, has less tourmaline, and hardly any of gem quality. The Riverside County localities, in the San Jacinto Mountains, include the Fano (originally the Simmons) mine, a rich producer of colored tourmaline, and the Columbia (also called the April Fool) mine, which was the first one discovered in the State, and has yielded many beautiful gems, but has not been worked much of late.

QUARTZ.

Quartz is one of the commonest of all minerals, occurring in rock-masses nearly pure, and forming a large proportion of most of the granitic rocks. Its hardness, and the fact that it is unaffected by most chemicals, render it very stable and persistent, and hence it forms the largest part of most sands and sandstones. Its varieties are almost innumerable in color and aspect, and many of them are beautiful as ornamental and semi-precious stones. The crystals of quartz are easily recognized by their peculiar form, that of a six-sided prism, long and slender or short and stout, terminated by a sharp six-sided pyramid at one or both extremities. Some of its varieties are the following:

Crystalline.	Non-Crystalline.
Rock-crystal (colorless).	Chalcedony (white and various pale tints).
Amethyst (purple).	Carnelian (pink to red).
Citrine (yellow).	Sard (dark red or brown-red).
Rose-quartz (pink).	Chrysoprase (green).
Smoky quartz (smoky).	Agate (banded, of various colors).
Cairngorm stone (smoky).	Rainbow agate.
Spanish topaz (deep yellow or brown).	Royal agate.
Morion (black).	Onyx (black and white, banded).
Plasma, prase (green).	Moss-agate, Mocha-stone.
Asteriated quartz.	Hydrolite (inclosing water).
Aventurine (spangled).	Jasper, an impure quartz (usually red, green or brown).
Quartz cat's-eye.	
Gold-quartz.	Bloodstone (heliotrope).
Dumortierite quartz.	Jasper agate (banded with different colors).
Tourmalinated quartz.	Lydian stone, Basanite, Touchstone (black jasper).
Hornblende in quartz.	
Göthite in quartz (Onegite).	Novaculite, Whetstone.
Rutilated quartz (Sagenite or Fléches d'amour).	Agatized wood.
	Jasperized wood.
Thetis hairstone.	

Many of these varieties exist in California.

Rock-crystal.—Transparent colorless quartz, or rock-crystal, though not rare as a mineral, is seldom found in masses of large size. When it is, however, it is valuable for use in the ornamental arts. One or two localities in the Alps, which have been known and worked from Roman times, though very difficult and perilous of access, have furnished material for the elegant carved objects to be seen in European palaces and museums.

In Japan also, large crystals were formerly obtained, from which were made the polished balls so much prized by the natives, and afterwards by foreigners, who have now almost drained the country of them by purchase. Japan has not yielded much new material for some years, and the main supply of it has been derived from Madagascar and Brazil. Within the last decade, however, very fine rock-crystal masses have been obtained in the United States, especially in California.

In 1891–92 an important development of crystallized quartz was made at Placerville, El Dorado County, by Mr. James Blakiston, in a quartz ledge running north and south, and dipping eastward about 45 to 50 degrees.* The rock of the ledge is partly decomposed and partly compact, and is traversed perhaps a hundred feet by a vein of crystallized quartz varying from 6 to 14 inches in width. This vein is also decomposed, and is filled in with a reddish earth or sand, and can be dug into with a stick or board. It is full of quartz crystals, of all sizes, from that of a man's finger up to remarkable dimensions, some of them weighing as much as 80 or 90 pounds. Several of these, over 50 pounds in weight, were pellucid and free from flaws; while others have peculiar interest from remarkable inclusions of chlorite, 3 to 5 millimeters in thickness, at several depths in the crystal—thus marking successive stages of crystal growth, and making very striking "phantoms," generally of green chlorite on white quartz layers. Of still greater interest, however, are other quartz crystals, 2 to 4 inches in length and half that in transverse diameter, containing at and near their centers inclusions resembling groups or clusters of dolomite or siderite crystals, cream-white to brown in color, and consisting of many rhombohedra from 2 to 4 millimeters in diameter. On breaking the specimens, however, the curious fact appears that these groups are hollow cavities in the quartz, the spaces being lined with a layer of chalcedony, or when brown, occupied only by a brown silicious material.

ILL. No. 17. Quartz Crystal from Placerville, El Dorado County. Weight 346 lbs. In Morgan collection, American Museum Natural History, New York City.

This would indicate that the original mineral must have been siderite or ankerite, afterwards covered by successive growths of the quartz, and in some manner decomposed during that process.

The most remarkable California quartz discovery, however, was made in 1897, by John E. Burton, of Milwaukee, Wis., in Calaveras County, at the old Green Mountain mine, in Chile Gulch, a mile and a half south of Mokelumne Hill. Here, in one of the ancient river channels, about 350 feet wide, filled with auriferous gravel and covered by an overflow of lava—which are characteristic of this portion of California—were found a quantity of enormous quartz crystals, embedded in the old

*Am. J. Sci. (3), 1892, Vol. XLIII, p. 329. Gems and Precious Stones of North America, Kunz, p. 351 (Appendix), 1892.

gravel. It is claimed that twelve tons were taken out in the years 1897 and 1898; one giant crystal, surrounded by an attached cluster of forty-seven smaller ones, weighed over a ton. A number of the finest were sent to New York, and splendid balls were cut from them by machinery constructed for the purpose. One of the largest crystals measured in the rough 19 by 15 by 14 inches, and another 14 by 14 by 9 inches. A perfect sphere cut from one of these crystals has a diameter of $5\frac{1}{4}$ inches, and is absolutely without a flaw. It is worth $3000. Even larger spheres have been cut from these California crystals—two ranging up to $7\frac{1}{5}$ inches in diameter, but they are not entirely flawless. The first mentioned sphere, and one of $7\frac{1}{5}$ inches, cut by Tiffany & Co., are the largest ever produced in this country, and are justly held among the most valued treasures of the Morgan-Tiffany collection; and one $7\frac{1}{2}$-inch ball is in the collection of the Musée d'Histoire Naturelle at Paris. A large, perfect hexagonal crystal in the American Museum of Natural History weighs 346 pounds. There were in all some twelve tons of crystals found, but few yielded cutting material.

Some of the crystals here found are apparently the largest ever obtained anywhere. The exteriors were frequently roughened or discolored, in other cases quite bright; but within, the quartz was clear and colorless. The great crystal in the one-ton group just mentioned showed a portion in its interior that was beautifully clear for a space of 14 by 16 by 24 inches, and might yield a flawless ball of over a foot in diameter. It was with great difficulty that this splendid mass could be taken out of the tunnel without injuring it. As the crystals show little wear, they have evidently not been transported far from their source, and must have come from some vein or fissure cut through by the ancient streams a little way above. If this could be discovered, it might be a most remarkable crystal mine.

So far as the gravel deposit here was explored by drifts, the crystals appeared to be strewn through it, and it would seem that there must be a large amount of valuable material there; but no further actual work has been done for several years.

Other California localities for crystals of several pounds in weight are Drum Valley, Three Rivers, and Yokohl, in the neighborhood of Visalia, in Tulare County.

Many fine crystals are also obtained in some of the mines in San Diego and Riverside counties, where they occur in the pegmatite veins that are worked for tourmaline and beryl. The Fano mine, in Riverside County, near Coahuila, has sold some 200 pounds of choice quartz crystals; they are also frequent at the Himalaya mine, at Mesa Grande, both colorless and smoky, at the Mack mine on Mount Palomar, and in most of the mines throughout that region of the State. Some of these may yield good material for use in the arts.

The Indians used quartz of all kinds for arrow- and spear-points. Those made of colorless transparent rock-crystal are rare and beautiful; some specimens of these, only an inch long, from Calaveras County, were presented to the State Mining Bureau, some years ago, by **Mr.** J. Z. Davis.

Amethyst, the purple variety of quartz, has not been found to any

ILL. No. 18. OUTCROP OF ROSE QUARTZ—FREE ZONE, NEAR THE MEXICAN LINE, SAN DIEGO COUNTY.

great extent in California; though it occurs somewhat in Mono County, at the Noonday mine, Bodie district.

Smoky Quartz is somewhat abundant, but has been already mentioned at the principal points, under Rock-crystal.

Rose-Quartz is a semi-crystalline translucent variety, of pale pink color, sometimes slightly opalescent. It is capable of use for many ornamental purposes being cut into balls, pendants, small vases, etc.

Rose-quartz is a substance much appreciated by the Orientals, and it has been successfully imitated by them by dipping white quartz rock, after heating, into an aniline solution; this process opens cracks in the quartz, which upon cooling absorbs the rose color, and retains it permanently.

Rose-quartz of choice pink color exists in some quantity at several points in Tulare County. Specimens from Yokohl and Three Rivers have been sold at remunerative prices; and according to Prof. W. H. Smith, of Visalia, it is found, of good quality, at several other places in that vicinity. Rose-quartz is frequent also in the pegmatite veins of the gem mines in San Diego and Riverside counties (described under beryl, tourmaline, and kunzite). It is also reported as a large outcropping ledge, rising above the surface of the ground for a distance of 240 feet, with a width of 40 feet, at a point near the Mexican boundary; the location is given by Mr. N. G. Douglas, who describes it as 45 miles from San Diego and 29 miles from Tia Juana, on the public road from the last-named place to Ensenada, in the so-called "Free Zone." It is claimed that the material can be quarried out here in blocks of any desired size. Another occurrence of rose-quartz, somewhat opalescent, is announced by Mr. Edward N. Walsh, at Escondido, San Diego County.

Gold-Quartz.—One of the most characteristic ornamental stones of California is the gold-quartz, which has been used to a large extent for jewelry and art objects. The gold of the placers and gravel-beds is derived from auriferous quartz veins in the metamorphic slates of the Sierra Nevada, having been worn out and washed down in the slow decay of those rocks. But the most important mining operations are now conducted upon the veins themselves, where the quartz is taken out and crushed. In most cases the gold, even in a rich vein, is scarcely visible; but sometimes it is so abundant and so conspicuous that it makes a strikingly ornamental stone—a matrix of white quartz, either opaque or translucent, through which gold is distributed freely in little patches or stringers. Jewelers pay from $20 to $30 for each ounce of gold contained in such material—the gold itself being worth about $16.50. The price of specimens varies from $3 to $40 an ounce, according to their beauty and to the proportion of gold included. This latter is determined by specific gravity.

Some crystals of limpid quartz, containing particles of native gold, have been found in California. One of these was said to have been an inch long, inclosing in the center a scale of gold about the size of the lunule of a finger nail. In Nevada County, in the Grass Valley mines, quartz is occasionally found supporting gold between the crystals. Most of the white gold-quartz comes from the counties of Butte, Calaveras, El Dorado, Mariposa, Nevada, Placer, Sierra, Tuolumne, and

Yuba. A black gold-quartz is found at the Sheep Ranch mine, in Calaveras County, and at Sutter Creek, in Amador. Occasionally bowlders of gold-quartz are met with. A smoothly rolled mass of this character was taken out of the bank of the Nevada Hydraulic Company, at Gibsonville. It weighed 160 pounds, and was judged to contain $2500 worth of gold; but its value for lapidary purposes was greater.

An artificial imitation of this stone has been made, by throwing grains of gold from crushed quartz into a fused milky glass, and allowing it to solidify in molds of various shapes. Another process was to produce a so-called rose gold-quartz, by backing a translucent gold-quartz with carmine paste. This was quite effective, especially when used in connection with the black or the opaque variety. The same thing might be done with backings of other colors, and a striking variety of effects produced. There occurs in Hungary an amethystine gold-quartz of great beauty, which could be readily imitated in this way.

Gold-quartz is made into a great variety of articles of jewelry and ornament, such as cane-heads, paper-weights, fan-sticks, bracelets, etc.; it is also employed for inlaying in certain kinds of elegant furniture, where it contrasts finely with dark wood or with other minerals of pronounced color, such as jasper, malachite, smoky quartz, etc.

The jewelry made from this material is mostly sold to tourists from the East and from foreign countries, though a good deal is also used in California. Eleven hundred dollars' worth was purchased some years ago, by an embassy from Asia; and almost all visitors buy specimens as souvenirs. The largest and most ambitious piece of work in gold-quartz is a model of the cathedral of Notre Dame in Paris; it is about a foot high, and is valued at $20,000. The amount of this material used in jewelry has varied greatly; it has at times been estimated as high as $40,000 a year, but has not approached such figures of late. One lapidary in Oakland, where most of the cutting is done, bought nearly $10,000 worth in one year, and a large San Francisco firm of jewelers bought nearly $15,000 worth. In 1902 the production amounted to only about $3000.

In the selection of the quartz for art work great care is necessary. The stones used must be large enough to bear the rough treatment of the diamond-saw and the lap-wheel of the polisher. All the rock quartz is friable, and some of it crumbles to pieces while undergoing these processes. The saw, catching in the gold in the slitting, prevents the cutting of large pieces, as the wafer-like slabs are apt to be broken by this resistance while being detached from the mass. For this reason, all the pieces set in cabinet work are small. Pieces 4 by 2 inches are quite rare, although fine pieces 4 inches square are at times seen. Rarely more than half of the rough material purchased finds its way into the market, owing to breakage while being trimmed into shape.

Quartz Inclusions.—Quartz is frequently penetrated by other minerals, and these combinations are often very beautiful and are valued as ornamental stones. Instances of this kind, mentioned in some of the writer's reports on the Mineral Resources of the United States, are the following: At the tourmaline locality near Coahuila, in Riverside County, the quartz is sometimes penetrated with fine hair-like crystals of tourmaline, so as much to resemble the beautiful rutilated quartz, or sagenite. One of the finest of these specimens is now in the collection of Harvard University, at Cambridge, Mass., and another in the American Museum of Natural History at New York.* A recent announcement has been made of the finding of smoky quartz penetrated by black and greenish-black tourmaline, near Fallbrook, 8 miles west of Pala, in San Diego County. Mr. W. H. Trenchard reports quartz inclusions of tourmaline, hornblende, and other minerals, as frequent throughout southern California, especially at the mines where the gem-tourmalines occur.

In Tulare County, on Deer Creek, Mr. L. B. Hawkins obtained specimens of a similar penetration of quartz by hornblende crystals.† This combination has also been noted by the State Mining Bureau, at Tyler's Ranch, near Oleta, Amador County, and at Fairplay, El Dorado County.‡

A notable occurrence of this character is that already described (p. 65) in El Dorado County, near Placerville, where a decomposed quartz vein is full of crystals ranging from the size of a man's finger up to 80 or 90 pounds. Several pieces of the quartz, over 50 pounds in weight, were pellucid and free from flaws, while others contained remarkable inclusions of green chlorite, 3 to 5 millimeters in thickness. Some of the crystals also contain inclusions of chalcedony, cream-white to brown in color.

Other interesting quartz inclusions are reported by Mr. H. F. Wheaton, of Riverside County, from the San Bernardino range, in the county of the same name, in the desert. These are the kind known by the name of sagenite, in which transparent quartz crystals are penetrated with long slender crystals of brilliant red or brown rutile, the oxide of titanium. This variety is well known and much admired, as both curious and elegant, and has received the name abroad of "fléches d'amour," or "love's arrows." When pieces of any considerable size are obtained, it is a beautiful material for carving into objects of art. There were also noted colorless quartz crystals with chloritic "phantoms" including "minute grouped masses of a green color," thought to be chrysocolla or epidote.

There are many other kinds of inclusions; and in some cases the included mineral so completely penetrates and fills the quartz as to

---

*Mineral Resources of the United States, 1893, p. 18.

† Ibid., 1897, p. 14.

‡ Ibid., 1894, p. 601.

change its aspect entirely, and render it almost opaque and highly colored. Such is the Mariposa County instance, referred to on page 97, under jade, where a quartz filled with a green micaceous mineral (mariposite) in minute scales forms a green stone so resembling jade that it is exported to China, where jade is greatly prized.

Dumortierite Quartz.—A very marked example of this character exists in Riverside County, on the Colorado Desert, about 50 miles north of Yuma and 11 miles west of the Colorado River. Here quartz is found that is so filled with a rare dark blue mineral, known as Dumortierite, that it closely resembles the celebrated stone long familiar in jewelry and the arts as lapis lazuli. Mr. John Stewart, of Los Angeles, who described the locality, states that the material is abundant and can be taken out in blocks of several hundred pounds, varying from light to dark blue and mingled blue and white. This should be a very fine ornamental stone, as it polishes beautifully. The locality, however, is remote and can only be worked in the winter or the rainy season, as water has to be hauled from the Colorado River, and the climate is too hot for either horses or white men from June to December.

Near Dehesa,* San Diego County, is a large body of lavender-colored dumortierite in quartz, which, if cut and polished, would make handsome specimens. This is evidently the mineral that was mistaken by the early collectors for erythrite, which it slightly resembles in structure and color. Erythrite has a metallic lustre, the dumortierite has none.

CHALCEDONY.

On some of the California beaches are found many interesting pebbles of chalcedony that frequently have the appearance of a partial polish. Among the most notable of these beaches are Crescent City, Pescadero, and Redondo. The first named is in Del Norte County, at the extreme northern end of the State. Pescadero is nearly central, lying some 25 miles directly west of San José, and may be reached by a beautiful mountain ride of 25 miles from San Mateo. The pebbles are found in great profusion, of different varieties of chalcedony and agate, many of them beautifully marked; others are apparently of jasper, and occasionally of fossil coral. Some of them are hollow geodes of chalcedony, inclosing a liquid with a moving bubble, like the "hydrolites" from Uruguay, South America, or from Tampa, Florida, and near Astoria, Oregon. These little sealed flasks, as they might be called, vary from the size of a pea, or less, up to rarely an inch in length, and are much sought after. Redondo is a favorite beach resort, about 15 miles south of Los Angeles. Here also are found many beautiful pebbles. It is

---

*Report of Dept. Min. Statistics, U. S. Geol. Survey.

quite the custom after each tide for the guests at the hotel to visit the beach to look for these treasures, which are especially abundant north of the pier. They are believed to come from a bed of sand and gravel in the vicinity. Several of these pebbles were found in 1901 in an Indian grave near Redondo.

At both these beach resorts, large quantities of the pebbles are gathered and sold to tourists, often in bottles of water, to bring out their

ILL. No. 19. PEBBLES FROM PESCADERO BEACH, SAN MATEO COUNTY.

varied colors. Long chains and other ornaments are also made by drilling the stones and stringing them on a flexible wire.

Another important pebble locality is that known as Moonstone Beach, on Santa Catalina island. The pebbles are not moonstones, but nodules of quartz weathered out of a rhyolite rock—composed of sanidine feldspar and quartz—while those of Redondo and Pescadero are agate and chalcedony, and come from amygdaloidal rocks.

Similar pebbles of varied coloring occur on Upper Spanish Creek, above Green Flat, in Meadow Valley, Plumas County, according to Mr. J. A. Edman.* Another locality is on the shores of Lake Tahoe, where

* U. S. Geol. Survey, Dept. Min. Statistics, 1899, p. 40 (reprint).

they are very abundant at points; and still another is Cañon Springs, in the southern part of the State, as announced by Mr. C. R. Orcutt, who found hydrolites and many beautiful agates in the drift of the desert and strewn over the mesas.*

Numerous handsome varieties of chalcedony are known to occur in California. Among these is a light blue variety, sometimes called sapphirine chalcedony. This is a stone that was highly valued in very ancient times, and was a favorite material for the carved Babylonian seals, 3000 to 4000 B. C. That used for this purpose came from Persia; it occurs also in the Urals, and at Treszytan in Hungary. Sapphirine chalcedony of equal beauty is found at Kane Springs, in Kern County, in masses of a deep sky-blue color, with the "botryoidal" or grape-

ILL. No. 20. PEBBLE BEACH, REDONDO, LOS ANGELES COUNTY.

cluster surface characteristic of much chalcedony. It does not seem to have been developed, however, nor is the amount of it to be had yet known.

Prof. William P. Blake mentions the occurrence of large masses of white chalcedony, delicately veined and in mammillary sheets, near the Panoche, in Fresno County, and also in Monterey County, and of pink chalcedony in nodules in the eruptive rocks in Los Angeles County, between Johnson's River and Williamson's Fork. There are doubtless many other localities in the State where handsome varieties of chalcedony occur, that may be developed and used. Some of the silicified wood, elsewhere noted as frequent in the old "sub-lava" gold gravels, is altered to white chalcedony and various types of agate.

* Report on Minerals of the Colorado Desert (10th Ann. Rept. State Mineralogist), 1890.

CHRYSOPRASE.

Chrysoprase is a chalcedony that is colored a beautiful light green by the oxide of nickel. It has been a favorite stone in jewelry and ornamental work from very ancient times, but is found at only a few places in the world. Within some years past several localities of chrysoprase have been opened in California, in the region about Visalia, in Tulare County. The first discovery was made as far back as 1878, by Mr. George W. Smith, a surveyor, who collected specimens of the mineral and submitted them to experts. Mr. Max Braverman, of Visalia, was the first who positively identified it as chrysoprase, from its content of nickel oxide. Specimens were thenceforth gathered and sent to various collections and museums. Later, its possible value for jewelry began to attract attention, and renewed search was instituted, resulting in the discovery of two more occurrences of it in the same county—one on Stokes Mountain and the other on Tule River. The original locality is at Venice Hill, about 12 miles northeast of Visalia. Here the chrysoprase forms small veins, 2 or 3 inches thick, in a jaspery rock. Much of it is flawed, but a good deal of choice material has been taken out and cut, and the color is good. The principal vein was located by Mr. C. P. Wilcomb, formerly of the Golden Gate Park Museum, San Francisco, while Mr. Braverman has been very active in searching and exploring. He and Mr. L. B. Hawkins located a fourth occurrence in 1897, in the same county, at Deer Creek, some 30 miles southeast of Visalia; and in the succeeding year a fifth locality was opened at Lindsay, half way between those two places. Here again the material is apt to be flawed, and much of it is pale in color; but a good deal has been taken out, and one remarkably fine specimen has been presented by Mr. Braverman to the State Mining Bureau. The stone here is mingled with a beautiful semi-opal of the same green colors, for which Mr. Braverman proposed the name chrysoprase opal or chrysopal.

There was considerable activity at these several mines for the first few years, but no large or steady production. Twelve hundred pounds were shipped from the Deer Creek locality, five hundred from that at Lindsay, and three hundred from Venice Hill; but the proportion of real gem-material was small. More recently, however, a New York company has taken up chrysoprase mining with more system and more success. Large amounts of fine material have been taken out, varying from deep to light green, and a great deal has been cut, in some cases into stones weighing several ounces each. Most of it, however, had been cut into squares, rhombs, ovals, etc., for studs, links, rings, and inlaying or mosaic work. The great increase in production is strikingly shown by the fact that the value of the annual output, which for several years before had been only about $100, leaped to $1500 in 1901, and to $15,000 in 1902.*

MOSS-AGATE (MOSS JASPER).

Chalcedony with dendritic markings, in masses from 15 to 18 inches across, and jaspery agate, with moss-like markings of a dark brown color, are among the minerals collected by Mr. H. F. Johnson in the San Bernardino Mountains, in the county of that name, in the desert region of California, and reported by Mr. Wheaton, of Palm Springs, in the adjacent county of Riverside.†

JASPER.

Jasper is also found at many points. It occurs of fine quality at Murphys, in Calaveras County, of various shades of red, brown, and yellow. Mr. J. A. Edman has described it in Plumas County, west of Meadow Valley, gray, green, red, and banded red and white; and some of these richly mingled varieties he thinks would be well adapted for use in the arts. Red and green jaspers are abundant near San Francisco, where an impure variety has been used for buildings and sidewalks. Around the shores of Lake Tahoe are scattered myriads of agate and jasper pebbles, which of course indicate the presence of these minerals in the rocks of the vicinity. A very peculiar variety occurs near San Francisco, as inclusions in basalt. The rock is made up of red spherulites consisting of quartz stained by iron, usually very small, but occasionally reaching a diameter of more than an inch. When cut and polished, this rock makes very handsome specimens.

OPAL.

H.= 6. G.= 2.0–2.2.

Opal is essentially silica, though differing from quartz in containing small amounts of water. It has numerous varieties, such as

Noble or precious opal,	Opaline,
Fire opal,	Hyalite,
Milky opal,	Hydrophane,
Opal agate,	Wood opal,
Moss opal,	Opalized wood.

Opal of different kinds has been observed at a number of points in California, but the precious or gem variety has not yet been obtained

---

*Mineral Resources of the United States, U. S. Geol. Survey, 1902, p. 81;—table of production of precious stones.

† Ibid., 1900, p. 37 (reprint).

of a size and quality to warrant mining operations. In 1897, fire opals, small but good, with larger pieces of inferior quality, were identified by the State Mining Bureau, from Dunsmuir, in Siskiyou County. Very recently a locality of precious opal that may prove important has been reported at the opposite end of the State, in the region of the Mojave Desert. Mr. C. R. Orcutt, of San Diego, describes the opal as occurring in large quantities in a porphyritic rock. Most of that which he had seen is chalcedonic, but some true precious opal has been found, and good stones have been cut from it. The locality is in San Bernardino County, some 25 miles northwest of Barstow. Here, opal of various colors, some almost amber-yellow, and some—though not very much—of the precious and fire opal, occurs in seams, veins, and pockets, in what is called an indurated clay, by a more recent describer, but is doubtless a decomposed igneous rock. The deposit is about half a mile wide and two miles long, dipping rather steeply, with an overlying tufa, and probably resting upon granite beneath. The locality is situated in a branch of Black's Cañon to the east of the main cañon. On the surface are found large deposits of what is called "semi-opal," a stone which resembles the genuine opal, but has not its beauty. The real opals are found at a little distance from the surface and resemble very much the Australian opals. J. C. Reed and associates, of San Bernardino, are the owners. This locality may prove valuable when more fully developed.

A beautiful yellow opal, without fire, and more resembling amber, was noted in 1895, by Mr. M. Braverman, at the chrysoprase locality at Yokohl, near Visalia, in Tulare County. This might make an attractive semi-precious stone, if procurable in any quantity. He also found a beautiful green variety, for which he proposed the name of chrysoprase opal, or chrysopal, at the chrysoprase mine near Lindsay, in the same county.

Rich white opals, but with no fire, were described as long ago as 1886, by Prof. William P. Blake, from near Mokelumne Hill, Calaveras County. They were found loose in a gravel stratum, 345 feet deep in a shaft sunk in Chile Gulch, in Stockton Hill; the pieces ranged up to the size of a walnut, and were at first supposed to have real value. A similar variety occurs with magnesite in Alameda County and at some other points, but is of little importance.

In San Bernardino County, milky opal was reported by Mr. Dwight Whiting, in 1897, in a narrow seam of sandstone, a little east of the N.E. ¼ of Sec. 24, T. 1 S., R. 13 W., San Bernardino base and meridian.

The peculiar, glassy, transparent variety of opal known as hyalite has been found in Lake County, in some abundance, by Mr. H. H. Myer, and is also reported as occurring with semi-opal, about 30 miles south of Mount Diablo, in the range of that name. It is an interesting variety, but not capable of any particular use in the arts.

Silicified and Opalized Wood.—Silicified wood, which is variously known as agatized wood and wood opal, is found in abundance in California. The ancient gold-bearing gravels, overlaid by a capping of lava or tufa, elsewhere described, frequently contain trunks and branches of trees that have been permeated by silicious waters and thus fossilized. In some cases the replacing material is opal silica; and very interesting gradations in the process of change can be seen, even in parts of the same specimen. Pieces from these "sub-lava" gravels from Nevada County (e. g.) will show parts completely silicified, but opaque and pale-colored, and others entirely altered to opal silica, of various shades of translucent brown. These logs and fragments represent trees that grew on the banks of the ancient streams, before the lava or tufa flowed into the valleys and filled them, and in some instances they are even found still standing erect.

Of the manner of this remarkable change, Prof. Joseph Le Conte* says: "In a good specimen of petrified wood not only the external form of the trunk, not only the general structure of the stem—pith, wood, and bark—not only the radiating silver-grain and the concentric rings of growth are discernible, but even the microscopic cellular structure of the wood and the exquisite sculpturings of the cell-walls themselves are perfectly preserved, so that the kind of wood may often be determined by the microscope with the utmost certainty, yet not one particle of the organic matter of the wood remains. It has been entirely replaced by mineral matter, usually some form of silica."

The general theory of petrifaction is as follows: When wood is soaked in a strong solution of iron sulphate (copperas), then dried, and the same process repeated until the wood is highly charged with this solution, and then burned, the structure of the wood will be preserved in the peroxide of iron that remains; also it is well known that the smallest fissures and cavities in rocks are speedily filled by infiltrating waters with mineral matters; hence wood buried in soil soaked with some petrifying material becomes highly charged with the same and the cells filled with the infiltrating material, so that when the wood decays the petrifying material is left, retaining the structure of the wood. Furthermore, as each particle of organic matter passes away by decay, a particle of mineral matter takes its place, until finally all of the organic matter is replaced. The process of petrifaction is therefore one of substitution as well as of interstitial filling. From the different nature of the process in the two cases, it happens that the interstitial filling always differs, either in chemical composition or in color, from the sub-

*Elements of Geology, p. 192.

stituting material. Thus the structure remains visible, although the mass is solid.

Agatized wood in large quantities, consisting of trees from 12 to 35 feet in length and from 18 inches to 2 feet in diameter, has been obtained near Calistoga, in Napa County; and in the hydraulic mines of California, at many points, large and very beautiful masses of opalized wood, of fine brown, yellow and black colors, are frequently found.

ALBITE—FELDSPAR.

Aventurine. Moonstone. Peristerite.

H.= 6.   G.= 2.62.   Luster vitreous, often pearly on cleavage surface. Color white, also bluish, grayish, reddish, and greenish; occasionally

ILL. No. 21. MOONSTONE BEACH, CATALINA ISLAND, LOS ANGELES COUNTY

having a bluish chatoyancy or play of colors.   Silica, 68.7; alumina, 19.5; soda, 11.8.

Albite (soda feldspar) is a constituent of many crystalline rocks, often associated with common or potash feldspar (orthoclase) in granite and pegmatite, and frequently in syenite, greenstone, and the crystalline schists.   The most common occurrences are in veins or cavities in granite or granitoid rocks.   Found in Calaveras, Inyo, Marin, Mono, Riverside, San Bernardino, and San Diego counties.

Both albite and orthoclase feldspar are especially conspicuous in the dikes or veins of pegmatite that traverse the granite and diorite rocks of San Diego and Riverside counties, and that carry the gem-minerals of that region.   They constitute the greater part of the pegmatite, together with quartz, and are often finely crystallized. The albite in some instances appears with the tourmalines, etc., in the pockets, in beautiful transparent crystals; and the orthoclase at times is found in crystals of great size; some of these at Mesa Grande are penetrated with fine prisms of rubellite or other colored tourmaline.

The use of albite for gem purposes is practically restricted to those kinds showing a bluish opalescence or play of colors, an aventurine effect, or a moonstone effect when cut *en cabochon.*

Peristerite is a whitish, adularia-like albite, presenting a bluish chatoyancy usually more or less mixed with pale green and yellow, the play of colors resembling that on the neck of a pigeon.

Aventurine is a grayish-white to reddish-gray albite, with internal fire-like reflections proceeding from minute inclusions, disseminated crystals.

Moonstone is a semi-transparent albite, having a chatoyant reflection resembling that of a cat's-eye, or an opaque pearly-white variety with a bluish opalescence. The so-called moonstones of Santa Catalina island, gathered on the " Moonstone Beach," are not albite, however, but nodules of translucent quartz. (See p. 72.) Minute crystals of the adularia variety of moonstone with beautiful blue reflections, occurring in a volcanic (rhyolite) rock, were found recently at Rialto, in the Funeral Mountains, in Inyo County, near the Nevada line. They are of wonderful beauty, but are valueless on account of their small size. They were supposed to be opals at first by many collectors who distributed them.*

It can hardly be said that any gem material of the species albite is known in California, although many interesting specimens have been found.

ORTHOCLASE—FELDSPAR.

H. = 6. G. = 2.5–2.7. Silica, 64.7; alumina, 18.4; potash, 16.9.

Orthoclase (potash feldspar) has already been referred to in connection with albite. Colors white to various shades of cream, gray, flesh-color and reddish, rarely green or brown. The only form in which it occurs in California, that is capable of ornamental use, is that of

Graphic Granite.—Graphic granite, or Hebrew Stone, appears at Pala, San Diego County, at the famous rubellite and lepidolite mine. Large masses of this peculiar rock, consisting of quartz and orthoclase so intergrown as to present the appearance of Hebrew writing, are found here as a phase of a large pegmatite dike. It is also found more or less at all the gem mines in the pegmatite veins of the San Diego-Riverside

---

* U. S. Geol. Survey, Min. Res. U. S., 1903, p. 44 (reprint).

district, and is known among the miners as "graphic-spar." When it exists in compact masses of uniform texture, it is capable of use as an ornamental stone, taking a fine polish, with delicate tints and curious pattern. Vases and other art objects that are very handsome are made of this material in Russia.

LABRADORITE.

H.= 6. G.= 2.72. Color gray, brown, or greenish. Labradorite is a lime-soda feldspar, containing silica, 53.1; alumina, 30.1; lime, 12.3; soda, 4.5.

The cleavable kinds often show a magnificent play of colors in which blue and green predominate, while yellow, red, pearl-gray, orange and amber are apparent. The mineral takes high polish and is then very handsome.

Observed in Mariposa County in Yosemite Park, and in San Bernardino County near Lytle Creek.

Mr. W. H. Trenchard, in a recent communication to the writer, states that labradorite occurs in the granitic intrusions of the entire gem-bearing district of southern California, with orthoclase and oligoclase feldspars.

DIOPSIDE.

H.= 6. G.= 3.2–3.38. Luster vitreous. Color ranging from white through several shades of green to dark green, and even nearly black. Silica, 55.6; lime, 25.9; magnesia, 18.5. It is a variety of pyroxene.

Diopside is occasionally cut as a gem. Pyroxene is a common mineral in serpentine and eruptive rocks; the variety diopside occurs in limestones and serpentines. It is observed in El Dorado County, in fine dark green crystals near Mud Springs; and in fine crystals at the Cosumnes copper mine.

ENSTATITE.

Bronzite. Hypersthene.

Enstatite.—H.= 5.5. G.= 3.10–3.13. Color yellowish, grayish, and greenish white. Luster vitreous or pearly. Silica, 60; magnesia, 40.

Bronzite.—H.= 5.5. G.= 3.1–3.3. Color grayish to olive-green and brown. Part of the magnesia of the enstatite is replaced by iron.

Hypersthene.—H.= 6. G.= 3.4–3.5. Color dark, brownish green, grayish black, greenish black, pinchbeck brown.

These minerals are all orthorhombic pyroxenes, and when cut *en cabochon* across the fibers, they all afford the cat's-eye effect. They are common constituents of peridotites and serpentines.

CONTRA COSTA COUNTY.—Bronzite at Mount Diablo.

SAN FRANCISCO COUNTY.—Bronzite at Fort Point and near to Orphan Asylum.

KUNZITE—SPODUMENE.*

H. $= 7-7\frac{1}{2}$. G. $= 3.19$. $SiO_2 = 64.5$; $Al_2O_3 = 27.4$; $Li_2O = 8.4$. Silicate of alumina and lithia.

A most remarkable discovery of transparent lilac-colored and pale pink to white spodumene has lately been made in California. The crystals were obtained close to a deposit of colored tourmaline, itself of notable interest, on Pala Chief Mountain, a mile and a half northeast from Pala, in San Diego County. This new discovery is but half a mile northeast from the celebrated rubellite and lepidolite locality† on Pala Mountain, where recent developments have brought to light immense quantities of amblygonite—this latter species occurring by the ton, while the lepidolite is estimated by the thousand tons. The locality is thus unequaled in the world for its abundance of lithia minerals.

Spodumene has long been known to mineralogists, but only within recent years has it been ranked among gem-minerals. It is a silicate of alumina and lithia, rather complex in constitution and peculiarly liable to alteration, the first effect of which is to destroy its transparency, so that most of the spodumene found is opaque and of little or no beauty. In this condition it is somewhat abundant at several localities in New England and also in Pennington County, South Dakota, the crystals often being very large, but dull and unattractive. It began to be recognized, however, some twenty-five years ago, that all these crystals had undergone alteration and must originally have been very beautiful. The change had proceeded from without inward, and at the center were found portions that still retained the color and transparency that once belonged to the whole. Even these remnants, however,

---

*(1) "A New Lilac-Colored Spodumene from Pala, California," by George Frederick Kunz, Am. J. Sci., Vol. XVI, September, 1903, pp. 264-267.

(2) "Kunzite: A New Gem," by Charles Baskerville, Science (N. S.), Vol. XVIII, September 4, 1903, pp. 303-304.

(3) "Spodumene from San Diego County, Calif.," by Waldemar T. Schaller. Bull. Dept. Geol., Univ. of California, Vol. III, No. 13, September, 1903, pp. 265-275.

(4) "The Lilac-Colored Spodumene (Kunzite) from California," Am. J. Sci., Vol. XVI, October, 1903, p. 335, with remarks on Action of Radium on Kunzite, by Sir William Crookes.

(5) George F. Kunz, N. Y. Acad. Sciences, "On the Action of Radium," etc.; January 4, 1904.

(6) "Gem-minerals of Southern California," G. F. Kunz, Science (N. S.), Vol. XIX, p. 472, January 15, 1904, pp. 107-108.

†G. F. Kunz, Mineral Resources of the United States, Washington, D. C. (reprint), 1893, p. 17; 1899, p. 38; 1901, p. 31.

were so fissured and marred that they could hardly be used for gems; but they indicated a lost elegance that led the writer to apply to spodumene the expression "a defunct gem." Since then, however, it has been found in the unaltered state and in several colors at two or three localities, and has come into recognition as an interesting and beautiful gem-stone.

The name spodumene is from the Greek *spodos*, ashes, from the dull whitish color of most of the altered crystals. In Europe the mineral is also frequently called triphane. A transparent yellow variety is known from Minas Geraes, Brazil, and these specimens have been to some extent cut into gems. In 1881, Mr. W. E. Hidden discovered numerous clear, bright green crystals at Stony Point, Alexander County, North Carolina, which were found in seeking for emeralds. Their real character was not recognized at first, and they were supposed to be cyanite or diopside; but an analysis by Dr. J. Lawrence Smith, of Louisville, Ky., showed them to be spodumene. He proposed for this new variety the name of hiddenite, which it has since borne, and it has also been called lithia emerald. This discovery excited much interest, and the new and beautiful American gem at once came into favor. The yield, however, was limited in amount, and for several years past, because of litigation and from other causes, the mine has not been worked.

The green hiddenite spodumenes, although extremely beautiful, were small; but the California crystals are of noble size, as well as very attractive in their color tones, varying from rich rosy lilac when found at some depth, to pale or almost colorless nearer the surface—doubtless due to weathering or the action of sunlight—in striking contrast to the rich deep pink-purple of those lower down.

These spodumenes are of remarkable size, transparency, and beauty. The following are the weights and dimensions of seven of the principal crystals:

	Weight, grams.	Weight, oz. troy.	Dimensions, centimeters.
No. 1	528.7	17.1	17 x 11 x 1
No. 2	528.7	17.1	22 x 8 x 1.5
No. 3	297.0	9.55	19 x 5.5 x 1.5
No. 4	256.6	8.25	23 x 4 x 2
No. 5	340.5	10.95	13 x 6 x 2.52
No. 6	239.5	7.70	18 x 4 x 2
No. 7	1000.0	31.0	18 x 8 x 3

They are extraordinary objects to the eye of the mineralogist. To see flat spodumenes of characteristic form, as large as a man's hand, but with bright luster and perfect transparency, and of this delicate pink-amethystine tint, is a novel and unlooked-for experience, as all the large crystals of this mineral ever before seen were dull grayish white, and perfectly opaque. At Branchville, Conn., some of the very large spodumene crystals, on being broken, were found to have remnants at their center, that retained some transparency and an amethystine color—just enough to indicate something of what their former beauty had been. The same mineral has now been found in its unaltered state, and the discovery is of great interest. Prof. Charles Baskerville, of the University of North Carolina, who has from the first been engaged in important studies on these specimens, has proposed for this lilac variety the name of kunzite,* after the writer, who was the first to determine it as spodumene.

The chemical composition of these specimens has been carefully determined by two independent authorities. The results appear in the analyses here given, of which No. I is by Dr. Waldemar T. Schaller,† of the U. S. Geological Survey, and No. II by Prof. Charles Baskerville and Mr. R. O. E. Davis, of the University of North Carolina:‡

	I.	II.
SiO_2	64.42	64.05
Al_2O_3	27.32	27.30
Mn_2O_3	.15	....
MnO	....	.11
CaO	....	.80
ZnO	....	.44
NiO	....	.06
Li_2O	7.20	6.88
Na_2O	.39	.30
K_2O	.03	.06
Ign.	....	.15
Totals	99.51	100.15

Occurrence.—Soon after the first announcement of this remarkable discovery, the locality was visited and examined by Dr. W. T. Schaller, then of the geological department of the University of California. In his report he describes the occurrence as follows:§

"The formation in which these fine crystals are found at the Pala locality consists of a pegmatite dike, dipping westerly at a low angle, perhaps 20°. It is more or less broken, and as a whole seems to form the surface of much of the slope of the hill on which it occurs. The dike is rather broad, but irregular in its present shape, and has a thickness probably of not more than thirty feet. So far as the mining developments have shown, only a small portion of the dike is rich in lithia minerals. Ordinarily, the dike is a coarse muscovite granite, the orthoclase and quartz predominating, containing many rounded prisms of black tourmaline, with broken ends. Lepidolite occasionally seems to replace the muscovite and when it does, red, blue, and green tourmalines replace the black variety. It is with these gem-tourmalines

* Science (N. S.), Vol. XVIII, No. 304, September 4, 1903, p. 303.

† Bull. Dept. Geol. Univ. Cal., Vol. III, Sept., 1903, pp. 265-275.

‡ Am. J. Sci. (4), Vol. XVIII, July, 1903.

§ Loc. cit.

that the spodumene occurs. While the tourmaline and lepidolite are frequently inclosed in the quartz and feldspar, no such inclusions of spodumene have been found. The latter mineral always occurs associated with the other minerals, but never penetrating them or penetrated by them. It occurs in pockets, and these facts seem to indicate that the formation of the spodumene is later and not coincident, in time of formation, with the tourmalines and with the dike. The dike cuts across the large intrusion of dark rock occurring at Pala and briefly mentioned by Dr. H. W. Fairbanks.* This large body of dark rock, several miles across, is surrounded on all sides by granite. The dark rock forming the foot-wall of the dike in which the spodumene occurs is a diorite, consisting of hornblende, a plagioclase, and (subordinate) orthoclase with accessory magnetite and apatite."

It will be seen by comparing this account with the general introductory statement already given, of the geological character of the San Diego-Riverside lithia and gem region, that the features here are those characteristic of most of the mines referred to, where gem-tourmalines and other rare and interesting species occur. The kunzite variety of spodumene is now found at several points in San Diego and Riverside counties in connection with the richly colored tourmalines. Although this particular locality on Pala Chief Mountain is the most prominent one, and has yielded most of the fine material, yet it was not the first at which kunzite was met with. Specimens of this mineral first reached the writer and were identified, in December, 1902, through Messrs. Tiffany & Co., from Mr. Frederick M. Sickler, who thought them perhaps a variety of tourmaline. Their exact locality was not given. In August, 1903, he announced that they came from the White Queen mine, near Pala. The crystals, though much smaller in size, are similar to those obtained soon afterwards from the Pala Chief.

Mr. Sickler, an explorer long familiar with the region, claims to have been the original discoverer of this mineral some years before, jointly with his father, Mr. M. M. Sickler; but its composition was not known, and from its association with tourmaline it was regarded as a peculiar variety of that species, although with some question if it were not a new mineral. Mr. Sickler visited Los Angeles in the summer of 1902, and tried in every way to find out from jewelers, mineralogists, and libraries, what the crystals might be, but without success. His discovery of the mineral, however, at the White Queen mine, on the ridge east of the Pala Chief, now known as Heriart Mountain, seems to be clearly entitled to priority.

During 1902 and 1903, much exploration and prospecting were carried on in this section by two Basque Frenchmen, Bernardo Heriart and Pedro Peiletch, as well as by the Messrs. Sickler, father and son;

---

*Eleventh Report State Mineralogist (Cal.).

and several claims were located by them, both jointly and separately, at points where kunzite and other interesting associated minerals were found. Most of these were on the eastern ridge, named after Bernardo Heriart; but the most important of all was the great kunzite

ILL. No 22. KUNZITE CRYSTAL, SICKLER MINES, HERIART MOUNTAIN, PALA, SAN DIEGO COUNTY.

and tourmaline mine on Pala Chief Mountain, which was located in May, 1903, by the two Frenchmen, with Messrs. John Giddens and Frank A. Salmons. The latter has been especially active in the subsequent development of this mine, which has furnished most of the fine kunzite that has thus far been placed on the market. The actual discovery of this kunzite mine must probably be credited to Heriart and Peiletch, early in 1903; although, as above stated, Mr. Sickler had found the mineral previously at the White Queen mine. It is also claimed that they and Mr. Sickler had located another kunzite claim some months before, on Heriart Mountain; this may be the same as the White Queen, or the Caterina.

On Heriart Mountain, which appears to be a foothill or spur of Agua Tibia Mountain, there are numerous dikes instead of one or two great ones, as on the Pala ridges. Many outcrops and openings show lepidolite, and several show kunzite at various points on the ends and on both sides of the ridge. Eleven claims are located and more or less developed. These are the following:*

San Pedro claim, north end; by Bernardo Heriart and Pedro Peiletch; lepidolite and gem-spodumene.

Sempe claim, crest and west slope; by the same; lepidolite, beryl, and colored tourmaline.

Anita claim, west side; Heriart and his partner; lepidolite.

Caterina claim, south side; Heriart and M. M. Sickler; lepidolite, amblygonite, and gem-spodumene at two openings.

El Molino claim, south side; F. M. and M. M. Sickler; gem-tourmalines.

Center Drive claim, south side; by the same; beryl and gem-tourmalines.

White Queen claim, south side; F. M. Sickler; lepidolite, beryl, and spodumene. This is the locality of the first kunzite crystals that reached the writer in 1902, as above mentioned.

Heriart claim, south and east side; M. M. and F. M. Sickler; lepidolite and gem-tourmalines.

Vanderburg claim, south slope; M. M. Sickler; lepidolite, beryl, gem-tourmaline, and gem-spodumene.

Naylor claim, east slope; F. M. and M. M. Sickler; lepidolite and gem-spodumene.

In addition to these the Sicklers, father and son, own the Fargo claim, on the west slope, which is promising, but hardly developed. They have recently reported the finding of a very fine, deep-colored crystal of kunzite, almost flawless, measuring 12.5 by 8 by 3 centimeters, at one of their newer claims on this mountain, 20 feet in the ledge and 16 feet from the surface.

In the coarse upper portion of all these pegmatite veins, a great variety of minerals have been developed. Mr. Sickler enumerates the following: Quartz—ordinary, milky, smoky, rose, and amethystine, also hyalite; orthoclase; albite; pyroxene; hornblende, green and black; spodumene—colorless, straw-yellow, lilac, and green; beryl—colorless,

* Min. Res. U. S., 1903, p. 942.

green, yellow, and rose; garnet; epidote; tourmaline—black and of many colors; micas—lepidolite, muscovite, biotite, damourite, and cookeite; montmorillonite; amblygonite; triphylite; and among the metallic oxides, hematite; sulphides, pyrite and molybdenite; bismuth, native and the oxide; also apatite, siderite, and columbite.

The peculiar reddish clayey substance, heretofore called montmorillonite, that occupies the cavities in which nearly all the gem-minerals are found, has recently been shown to be really a form of halloysite, by Dr. W. T. Schaller.*

Kunzite has also been found in Riverside County, at the Fano mine, near Coahuila, which was located in 1902 by Mr. Bert Simmons, and for some time bore his name. It occurs here both pink and colorless, chiefly, also some reported as yellow, green, and blue, associated with tourmalines of fine deep blue, green, and other shades, beryl, quartz crystals, lepidolite and amblygonite. The location is about 20 miles northeast from Pala, on a spur of the San Jacinto range, in Section 33, T. 6 S., R. 2 E., S. B. M.

The White Queen crystals (those that first came to the writer)* are usually smaller than those from Pala Chief, sometimes perfectly colorless, and varying from half an inch or less to two inches in length by one inch in breadth. Some are elegant specimens and some could be cut into gems. The hardness is about 7. They are perfectly transparent and remarkably free from flaws, and possess the spodumene pleochroism very markedly. Looked at transversely, the lilac ones are nearly colorless, or faintly pink; but longitudinally they present a rich pale lavender color, almost amethystine. The characteristic etching is also well developed, especially on the pyramidal faces; but the surfaces are dull, and are etched all over as if with a solvent.

Two crystals, the largest and another one, from the first lot received, gave the following measurements:

> (a) 53 mm. (2⅛ in.) and 35 mm. (1⅜ in.).
> (b) 37 mm. (1½ in.) and 27 mm. (1₁₆ in.).
> (c) 11 mm. (₇₁₆ in.) and 15 mm. (₁₁₆ in.).

The specific gravity determined on three crystals was found to be 3.183.†

	Weight, grains.	Specific Gravity.
Spodumene: Lavender	20.393	3.179
Yellow-white	8.359	3.185
Lavender	10.872	3.187

Since then, larger crystals have been found, some comparable to those from the Pala Chief, one in particular weighing 24 ounces, and many gems from this district have been cut in San Diego. The product of the Pala Chief mine is all sent to New York.

---

* Am. J. Sci. (4), Vol. XVII, March, 1904, p. 191.
† Min. Res. U. S., 1903, pp. 939, 940.

By a mere coincidence, just at the time when this mineral had come into notice, in the summer and autumn of 1903, an extended investigation on certain optical properties of the gem-minerals in the Tiffany-Morgan and Bement collections in the American Museum of Natural History, New York, had been undertaken jointly by the writer and Prof. Charles Baskerville, of the University of North Carolina, and was then in progress. This new mineral, just announced, proved to have specially marked and interesting peculiarities in connection with these experiments. Dr. Baskerville published a paper in relation to his observations, in the first of which he proposed to name this variety of spodumene after the writer, and described the special properties which he had observed in it.* In a second paper is given a fuller account of the subject, with the results of further studies, conducted with the writer.†

From this we quote: "In a recent investigation made by us on the behavior of a large number of minerals and gems with various forms of radiant energy, including the emanations, as well as on the production of luminescence in some cases by other physical means, the new variety of spodumene, designated kunzite, was found to be peculiarly sensitive, and to exhibit some remarkable properties.

"In general, as shown by these investigations, the gem-minerals were little affected by ultra-violet rays; but three species exhibited a high degree of responsiveness to these and to all forms of radio-activity, so far experimented with. These minerals were diamonds of certain kinds: willemite (zinc orthosilicate), which in some cases has been used as a gem-stone, and kunzite. The behavior of the last, as noted in various experiments, is unique and will be briefly described here by itself.

"1. *Attrition and Heat.*—Kunzite does not become luminous by attrition, or rubbing. Several specimens were held on a revolving buff cloth making three thousand revolutions per minute, so hot as to be almost unbearable to the hand, and still it failed to become luminous. Wollastonite, willemite, and pectolite are, however, very tribo-luminescent.

"As to luminescence induced by heat alone, it was found that kunzite does possess the property of thermo-luminescence to some extent, with an orange tint and at a low degree of heat."

It may here be noted that an observation made by Dr. W. T. Schaller indicates that kunzite does clearly show tribo-luminescence in certain cases.‡ He states particularly, as an interesting observation, that on cutting a crystal with a diamond saw, it became thoroughly luminous. This result may perhaps have been partly electrical.

*Science (N. S.), Vol. XVIII, 1903, pp. 304 and 769.
†Am. J. Sci. (4), Vol. XVIII, July, 1904, pp. 26-28.
‡Bull. Dept. Geol. Univ. Calif., Vol. III, Sept., 1903, pp. 265-275.

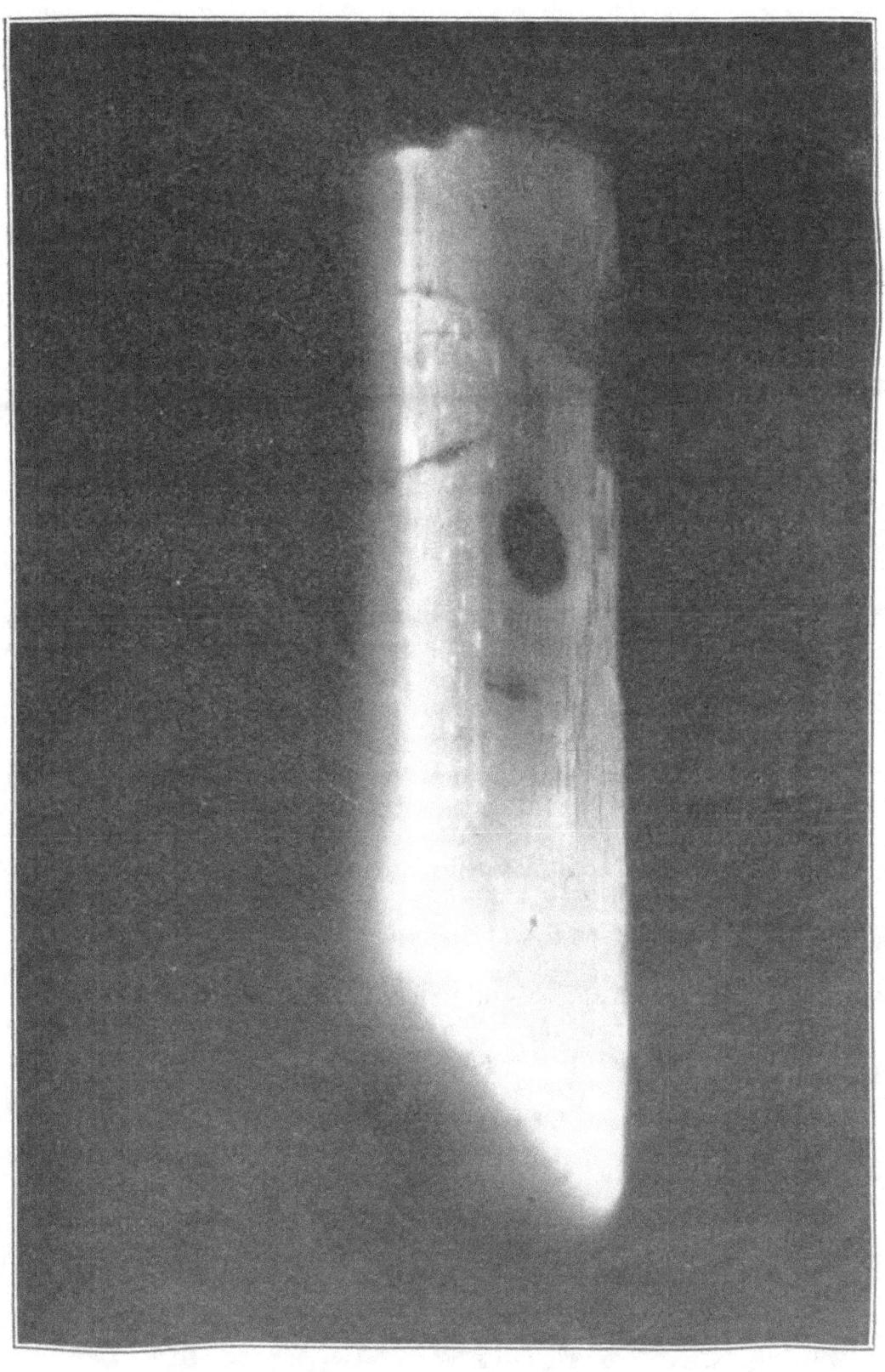

ILL. No. 23. AUTOPRINT FROM CRYSTAL OF KUNZITE EXPOSED TO THE X-RAY FOR
10 MINUTES, AND EXPOSED TO PLATE FOR 5 MINUTES (SHOWING THE
EMANATIONS). (Taken February 5, 1903, by Dr. H. G. Piffard.)

Continuing from the joint paper:

"2. *Electricity.*—The mineral assumes a static charge of electricity, like topaz, when rubbed with a woolen cloth. On exposing kunzite crystals of different sizes to the passage of an oscillating current obtained from large Helmholtz machines, the entire crystal glowed an orange-pink, temporarily losing the lilac color. A well-defined, brilliant line of light appeared through the center, apparently in the path of the current. On discontinuing the current, the crystal gave the appearance of a glowing coal. It was not hot, however, and the phosphorescence lasted for forty-five minutes.

"Three large crystals, weighing 200, 300 and 400 grams each, were attached to copper wires so that the current passed in one instance from below up, and from the other upwards across the crystal—first across the prism, then parallel with the prism. In each instance the crystals became distinctly luminous, a pale orange-pink, and between the two wires a bright almost transparent line passed from one wire to the other; in reality, as if the two elongated cones crossed each other, the line of the path being transparent at the sides, whereas the rest of the crystals appeared translucent. After the exposure of two minutes, they were laid upon photographic plates and in five minutes produced a fine autoprint, herewith shown. The crystals continued to glow for forty-five minutes.

"When a cut gem is suspended between the two poles, it becomes an intense orange-pink color, glowing with wonderful brilliancy. The discharge seemed as if it would tear the gem asunder, although actually it was unaffected.

"3. *Ultra-violet Rays.*—These invisible rays, produced by sparking a high voltage current between iron terminals, caused large crystals of kunzite, white, pink, or lilac, to phosphoresce for some minutes. The white responded most readily.

"4. *Roentgen, or X, Rays.*—All forms of kunzite become strongly phosphorescent under these rays. An exposure of half a minute caused three cut gems to glow first a golden pink, and then white for ten minutes. The glow was visible through two thicknesses of white paper, which was held over it. A large crystal (2 x 4 x 10 cm.) excited for five minutes, afterwards affected a sensitive photographic plate and produced an autoprint: it rested directly on the plate, separated by thin white paper, and remained for ten minutes in a dark room inclosed in a black box.* Another crystal, exposed for ten minutes, was laid for five minutes on a sensitive plate.† The resulting auto-photograph was clear and distinct, but presented a very curious aspect not seen by the

---

*Science (N. S.), Vol. XVIII, 1903, p. 303.

† This was made by Dr. H. G. Piffard, of New York City, and is shown of natural size in the accompanying plate. (See page 89.)

eye—as of a misty or feathery outflow from the side and termination of the crystal, suggesting an actual picture of the invisible lines of force. The other varieties of spodumene, both mineral material and cut gems, failed to show this property.

"Although kunzite is so responsive and fluorescent and so beautiful upon exposure to the X-ray, it is, however, like all silicates, opaque to the ray itself. Four crystals weighing 100, 200, and 400 grams each, were exposed to the Roentgen ray for two minutes. They became first a beautiful rose-orange, then assumed a white phosphorescence, and at the end of forty-five minutes there was still a faint residual glow. Two minutes' exposure to the X-ray caused them to print a perfect autotype. The glow in all instances showed first a rose-orange color, then a pale pink, finally resolving into a white fluorescence; the autoprint shows the feathery outlines of light or energy thrown out by the crystal.

"5. *Conduct with Radium Preparations.*—Exposed for a few minutes to radium bromide with a radio-active strength of 300,000 and 1,800,000 (uranium being taken as unity), the mineral becomes wonderfully phosphorescent, the glow continuing persistently after the removal of the source of excitation. The bromide was confined in glass. Six hundred grams of kunzite crystals were thus excited with 127 milligrams of the radium bromide in five minutes. The effect is not produced instantaneously but is cumulative, and after a few moments' exposure the mineral begins to glow, and its phosphorescence is pronounced after the removal of the radio-active body. The luminosity continued in the dark for some little time after the radium was taken away. No other varieties of spodumene examined, including hiddenite, gave like results. In this respect, as with the Roentgen ray, the kunzite variety stands by itself.

"When pulverized kunzite is mixed with radium-barium chloride of 240 activity or carbonate of lower activity, the mixed powder becomes luminous and apparently remains so permanently; *i. e.*, in several months no loss has been observed. The same is the case if pulverized wollastonite or pectolite be used instead of the kunzite. When either of these mixtures is put in a Bologna flask and laid on a heated metal plate (less than red-hot), the powder becomes incandescent and remains so for a long time after removal.

"These three minerals phosphoresce by heat alone, as was mentioned above in regard to kunzite. Perhaps this luminosity of the mixed powders at the ordinary temperature may be accounted for in part by the evolution of heat* on the part of the radium compounds, but there are experimental reasons which cause us to reject such explanation for the total effect.

"The *emanations* of radium, the *alpha*-rays, according to Rutherford,† are condensed at a temperature of — 130° to — 140° C. The

· * P. Curie and Laborde, Comptes Rend., CXXXVI, 673.   † Phil. Mag., V. 561.

emanations were driven from radium chloride by heat and condensed with liquid air on a number of kunzite crystals, and *no phosphorescence was observed.*   Consequently *kunzite responds only to the gamma-rays,* which are believed to be virtually Roentgen rays.

"6. *Actinium.*—A sample of the still more rare and novel substance discovered by Professor Debierne* and received from him through the courtesy of Professor Curie, was also tried as to its action upon kunzite and some other minerals.   The actinium oxide, with an activity of 10,000 according to the uranium standard, gave off profuse emanations and affected diamonds, kunzite, and willemite in a manner similar to the radium salts, with quite as much after-continuance.   However, we have not tried the condensation of these emanations upon the minerals by refrigeration.

"The peculiar properties of the kunzite variety of spodumene, which have been enumerated, have not been observed in any other of the gems or gem-minerals that we have examined.   It is barely possible that the small content of manganese may have much to do with it, but from our present knowledge basing a chemical explanation thereon is idle."

Prof. William Crookes, the eminent English physicist, conducted some similar experiments on the behavior of kunzite with radium bromide and obtained identical results, as stated by him in a letter to the writer in October, 1903.

Magnificent series of the finest crystals and large cut gems of kunzite are in the Morgan collection in the American Museum of Natural History (see plate), and in the Morgan collection in the Musée d'Histoire Naturelle, Paris.   A fine cut gem and fine crystals are also in the British Museum, London.

Use of Kunzite in Jewelry.—Kunzite has now been cut and sold as a gem for more than a year, and has been received with much favor as a new and a wholly American gem.   At first it was feared that it might be difficult to cut, as many specimens, being mistaken for a variety of tourmaline, were ruined in the attempt to cut them, because of their strong tendency to cleavage.   But the fact that kunzite spodumene has a facile cleavage in one direction, as have topaz and diamond, was soon understood by lapidaries who were familiar with the cutting of the hiddenite variety or of the yellow spodumene and topaz from Brazil.

The result is that there has been no difficulty in having the gem cut into every form — brilliant, degree-top, mixed brilliant, and other styles—and of sizes weighing from 1 to 150 carats each.   In color they vary from almost white with a faint pink tone through pink and lilac pink into dark lilac.   An English writer has called it "peach-blossom." The gem is remarkably brilliant, no matter what the color.   It is usually

---

* Compt. Rend., CXXIX, 593.

perfect and free from flaws, and, when pink, is one of the few natural stones of that color. As a lilac gem it is quite unique. The price has varied from $6 to $20 per carat, although generally it has averaged one third of the latter figure.

VESUVIANITE.

Idocrase.   Californite.

H.= 6.5.   G.= 3.35–3.45.   Color brown to green in various shades. A calcium alumina silicate. Fuses easily to a brown glass.

Californite (Vesuvianite).*—A discovery has been recently made of a mineral which promises well as an addition to the increasing list of semi-precious or ornamental stones found in the United States. It is not indeed a new mineral species, but a compact massive variety of vesuvianite (idocrase). It was first announced in the report of the U. S. Geological Survey for 1901, by the writer,[†] as having been found by Dr. A. E. Heighway, on land owned by him on the south fork of Indian Creek, 12 miles from Happy Camp and 90 miles from Yreka, in Siskiyou County. Here a hard and handsome stone, varying from olive to almost grass-green, and taking a fine polish, outcrops for some 200 feet along a hillside about 100 feet above the creek, and large masses have fallen into the bed of the creek below. It was at first supposed to be jade (nephrite), but proves upon analysis to be vesuvianite. The fallen pieces were in some cases as much as five feet square and two feet thick, of excellent quality for polishing, and of varying shades of light to dark green. The associated rock is precious serpentine.

This substance closely resembles a mineral from two localities in the Alps, on the south side of the Piz Longhin, in the Bergellthal, and found in rolled pieces in the bed of the stream called the Ordlegna, near Casaccia, in the Upper Engadine. These were at first taken for jadeite,[‡] but were positively identified as vesuvianite by the analysis of Berwerth.[§] It seems at first remarkable that the same mistake should have been made in both cases as to this massive vesuvianite, but its whole aspect is so jade-like that it is not surprising. The rich translucent green color, fine-grained sub-splintery fracture, and brilliant luster when polished all strongly suggest jade. The polished surface shows minute pale streaks or flocculi, which still further heighten the resemblance.

The following analysis was made by Mr. George Steiger, through Prof. F. W. Clarke, Chief Chemist of the U. S. Geological Survey, in the spring of the present year:

*N. Y. Acad. Sciences, October 19, 1903; N. Y. Min. Club, October 20, 1903.
† Mineral Resources of the United States (extract), 1901, p. 30.
‡ Fellenberg, Jahrb. Min., Vol. I, 1889, p. 103.
§ Ann. Mus. Wien, Vol. IV, 1889, p. 87.

Analysis of Vesuvianite, from Siskiyou County, California.

SiO_2	35.85
Al_2O_3	18.35
CaO	33.51
Fe_2O_3	1.67
FeO	.39
MgO	5.43
MnO	.05
TiO_2	.10
P_2O_5	.02
H_2O (below 100° C.)	.29
H_2O (above 100° C.)	4.18
	99.84

The analysis is essentially that of a normal vesuvianite, though the percentage of water is unusually high; the lime and the iron are below the average; the titanium and phosphorus are exceptional occurrences.

The mineral is compact, extremely tough, and readily takes a high polish, quite as beautiful as that of nephrite (jade), with which it was at first confounded. The hardness is 6.5, and the specific gravity (from two determinations), 3.286. The luster is vitreous, often inclining to resinous, and the streak white. The color is a yellow leek-green, with inclusions of a darker green, generally more translucent than the surrounding mass.

What appears to be the same mineral has recently been announced from two other localities quite remote from the first. One of these was reported by that indefatigable prospector, Mr. Max Braverman, of Visalia, as existing in Burro Valley, in Fresno County, a mile and a half from Hawkins school house, and 32 miles east of Fresno City. The material is pale olive-green, translucent, with darker spots in a paler mass. It breaks with an uneven fracture, slightly splintery and partly crystalline, and hence much resembles the Siskiyou County material.

The other locality is apparently not far from the last mentioned; it is said to be in Tulare County, near the town of Selma, which, though in Fresno County, is near the Tulare line. Here the mineral is of richer color, at times resembling the tint of apple-green chrysoprase, for which it was at first mistaken.

The following analysis was made of this material by Mr. George Steiger, through Prof. F. W. Clarke, Chief Chemist of the U. S. Geological Survey:

Analysis of "Californite" from Fresno County, California.

SiO_2	36.55
Al_2O_3	18.89
CaO	35.97
Fe_2O_3	.74
FeO	.74
MgO	2.33
CO_2	.91
F	.13
H_2O (below 100° C.)	.58
H_2O (above 100° C.)	3.42
	100.26

It will be seen on comparing this analysis with that of the Siskiyou County mineral, that they accord quite closely in essentials. In this case the carbon dioxide and fluorine are exceptional and doubtless due to impurities, as the titanic and phosphoric oxides were in the other case.

This interesting mineral exists in large quantity, and could be cut into a variety of ornaments, in the same way as jade, nephrite, and chrysoprase. It is a form of vesuvianite distinctive enough to receive a special variety name—which if appropriate and euphonious, would undoubtedly aid the sale of the stone in the arts. I have therefore proposed for it the name of *"Californite."*

Vesuvianite of gem quality, that is to say, in transparent crystals, was found by Mr. W. H. Trenchard as occurring near Jacumba, San Vicente, and some other points in that part of San Diego County.

PECTOLITE.

H.= 5. G.= 2.7–2.8. Color white. Usually fibrous radiated, also compact massive. Silica, 54.2; lime, 33.8; soda, 9.3; water, 2.7.

During 1887 a massive white pectolite of unusually dense structure, and susceptible of a high polish, was announced by William P. Blake as occurring in Tehama County, Cal., in masses of considerable size. In a letter to the writer he gives the following description of it:* "It occurs in a vein, and is broken out in rough tabular masses, from two to three or more inches in thickness, but it is reported that much larger masses can be obtained. It is exceedingly tough and hard to break. The fractured surfaces are irregular, without cleavage, but have a silky luster, and a crypto-crystalline structure is exhibited in extremely fine inseparable fibres, which are radial, curved, and interlaced, and are, perhaps, embedded in a siliceous magma, but the fibres constitute the bulk of the mass. Color white, with a delicate shade of sea-green; translucent. Exposed or weathered portions lose their porcelain-like translucency, and become white and somewhat earthy in appearance, and exhibit the crypto-fibrous structure with more distinctness. Specimens cut and polished across the end of a slab-like mass show on one side a narrow selvage of breccia made up of fragments of the pectolite and of a dark-colored rock, mixed and firmly cemented together. On the opposite side or border of the mass there are distinctly formed parallel planes of concentric layering, from the surfaces of which the fibres diverge. These layers and the brecciated border opposite show the vein-like formation of the mass between walls. Its hardness is from 6 to 6.5. It may be found useful as an ornamental stone for

---

*George F. Kunz, "Gems and Precious Stones of North America," 1890, p. 178.

making small objects, cups, plates, handles, or for carving figures, or inlaid work."

Massive pectolite, similar to the above, is also found at Fort Point San Francisco, in veins reaching several inches in diameter *

AXINITE.

H.== 6.5–7.   G.== 3.3.   Luster highly vitreous or glassy.   Color clove-brown, plum-blue, violet, pearl-gray, honey-yellow and greenish yellow. Strongly pleochroic.

A complex silicate of alumina, calcium, and manganese, with some iron and magnesia, and containing also from 5 to 6 per cent of boric acid and $1\frac{1}{2}$ per cent of water.   Occasionally cut for ornamental purposes.

INYO AND SAN BERNARDINO COUNTIES.—Axinite was found in several places in Death Valley, in the Funeral Range and Owl Mountains, by the State Mining Bureau expedition of 1902.

SAN DIEGO COUNTY.—Quite recently (1904) a locality has been discovered by Mr. Thomas A. Freeman, near Bonsall, where axinite crystals occur that are of remarkable beauty; in color they are a smoky pink, or "ashes of roses" tint, brilliant and perfectly transparent, and would cut into attractive gems.   The crystals are stated to be quite abundant, occurring in pockets, with crystallized quartz.   The locality lies about 18 miles south of Pala and 20 miles west of Julian.   No work has been done upon it lately; but as axinite is a rare mineral, the occurrence is very interesting, and it should be further developed.

JADE.

A general term applied to various mineral substances of tough, compact texture, and from nearly white to a dark green color, and even nearly black.   Properly, its use should be confined to *Nephrite* and *Jadeite*, but it is often applied to certain forms of minerals more or less resembling these, such as sillimanite, pectolite, serpentine, vesuvianite.

Jadeite.—H.== 6.5–7.   G.== 3.3.   Silicate of sodium and aluminum. Color apple-green to emerald-green, bluish green, leek-green, greenish white to nearly white.   Extremely tough.   Also known as *Chloromelanite*, when dark.

Nephrite (or Kidney stone).—H.== 6–6.5.   G.== 2.9–3.1.   Color from white to dark green.   A variety of actinolite or hornblende.   Tough and compact.

---

* Mineralogical Notes by A. S. Eakle.   Bull. Dept. Geol. Univ. of Cal., Vol. II, No. 10.

The lack of brilliancy makes it of little value for jewelry except as bracelets, but its great toughness renders it eminently suitable for ornamental and carved work, displaying delicacy of workmanship admirably.

MARIPOSA COUNTY.—Hacked quartz, a peculiar quartz which has a resinous luster and contains dense inclusions of the mineral *mariposite* (a green mica), has been shipped to China in considerable quantities, the pleasing green color making it a good substitute for the jade so highly prized by the Chinese.

DATOLITE.

H. = 5.   G. = 2.98.   Color white, creamy, grayish, pale green, yellowish, reddish or amethystine.   In small glassy crystals to massive. Silica, 37.6; boric acid, 21.8; lime, 35.0; water, 5.6.

This is another of the mineralogical gems of considerable interest, as it exists in a number of places in the desert portions of the State.

INYO AND SAN BERNARDINO COUNTIES.—Datolite of a creamy color, and pale green, is found in the borax districts of both of these counties, notably in Death Valley and at various points along the Amargosa River; in the Slate range, and in the Calico Mountains.   A specimen from San Carlos, Inyo County, associated with grossularite, is in the State Mining Bureau, No. 2190.

Crystallized datolite, as well as the compact massive variety, occurs at Fort Point, San Francisco.*

APOPHYLLITE.

Fish-eye Stone.   Ichthyophthalmite.

H. = 5 or less.   G. = 2.33.   Color from white to gray; occasionally tinged with green, pink or yellow.   Luster pearly to vitreous.   Silica, 53.7; lime, 25.0; potash, 5.2; water, 16.1; and occasionally a small amount of fluorine.

Occasionally cut for gem purposes.   Its pearly luster, producing an effect like that of a fish's eye, gave rise to the name "fish-eye stone."

Occurs in basalt at Buckeye mine, near Orion Valley, Plumas County; New Almaden quicksilver mines, Santa Clara County, and at Fort Point, San Francisco County, some of the crystals being replaced by silica.

* Bull. Dept. Geol. Univ. of California, Vol. II, No. 10.

LAPIS LAZULI.

Lazurite. Native Ultramarine.

H.= 5.5. G.= 2.38–2.4. Color rich Berlin blue, azure blue, violet blue, greenish blue. Translucent. Silica, 31.7; alumina, 26.9; soda, 27.3; sulphur, 16.9. Occurs commonly massive in limestone and in granite rock.

Lapis lazuli was long thought to be a simple mineral, but it consists of a mixture of a bluish substance called lazurite, with granular calcite, scapolite, diopside, amphibole, mica, pyrite, etc.

The richly colored kinds are highly esteemed for costly vases, mosaics and ornamental work.

MONO COUNTY.—Near Mono Lake.

Lazulite, or **False Lapis Lazuli,** is a rather complex phosphate of alumina and magnesia, with some water. H.= 5.6. G.= 3. Color azure blue.

SAN DIEGO COUNTY.—The first specimen of this mineral noted in California was found near Oceanside in 1893. S. M. B. 13591.

ANDALUSITE.

Chiastolite.   Macle.

H.= 7.5. G.= 3.2. Luster vitreous, often dull. Brittle; cleavage prismatic and distinct. Orthorhombic. Color varies from reddish or greenish brown to olive-green, flesh-red, rose-red, violet and pearl-gray. Pleochroism strong in some colored varieties, green in one direction, and hyacinth to rose-red in another. The variety Chiastolite or Macle varies in hardness from 5 to 7.5. Silica, 37; alumina, 63. Transparent to opaque.

Chiastolite or **Macle** is a variety of andalusite, the stout crystals having the axis and angles of a different color from the rest, hence exhibiting a colored cross or a tessellated appearance in a transverse section. These curious cross-like markings make it a favorite gem abroad, although there is but little demand for it in this country. It is often sold by jewelers under the name of "cross-stone."

FRESNO COUNTY.—The specimen No. 8747 in the museum of the State Mining Bureau is from this county; locality unknown.

MADERA COUNTY.—Location unknown. Specimens shown are of fine quality and remarkable size. Fragments of crystals found by W. W. Jefferis are over 3 inches long and measure 1¾ by 1¼ inches in diameter on the ends, the section being a rhombic prism. When polished, the stones show the peculiar cross-pattern in a rich black upon a white or fine salmon-colored ground, and sometimes with a black square or lozenge at the center, from which the arms of the cross extend.

MARIPOSA COUNTY.—Choice crystals have been found in the placers along the Chowchilla River, near the old road to Fort Miller, the crystals showing fine black crosses on a white ground in a remarkably perfect manner. They are also found in the conglomerates that cap the hills in that vicinity and probably come from the slates and schists a little higher up the river. Specimen No. 10756, State Mining Bureau, has a matrix of argillaceous schist, with mica. Small and imperfect macles have been found in the slates at Muller's ranch near Hornitos. S. M. B. 13573.

RIVERSIDE COUNTY.—Specimens of opaque pink andalusite from Coahuila.

EPIDOTE.

Including Thulite, Allanite, Zoisite.

Epidote.—H.= 6–7. G.= 3.2–5. Luster vitreous to resinous. Color peculiar yellow-green, to red, yellow, gray and colorless. Doubly refracting and strongly pleochroic, showing a green, a brown, and a yellow as viewed from the several directions. Silica, 38; alumina, 22; iron, 15; lime, 23; water 2, but variable.

Thulite.—H.= 6.5. G.= 3.3. Color peach-blossom red to rose-red. Strongly pleochroic, rose to yellow.

Allanite.—H.= 5.5–6. G.= 3.0–4.2. Crystals broadly tabular, or long acicular. Color nearly black.

Zoisite.—H.= 6–6.5. G.= 3.25–3.37. Color gray, yellowish brown, greenish gray, apple-green.

Epidotes are common in gneiss, schist and serpentine, and in sandstones adjoining trap rocks, in crystalline limestone.

CALAVERAS COUNTY.—Epidote crystals on quartz from Bald Point, Mokelumne River. S. M. B. 11856.

LAKE COUNTY.—Zoisite associated with glaucophane at Sulphur Banks.

MARIPOSA COUNTY.—At Mount Hoffman, epidote. S. M. B. 12006.

MADERA COUNTY.—From Grub Gulch. S. M. B. 13525.

SAN DIEGO COUNTY.—Found in clear transparent crystals at the McFall mine, 7½ miles southwest of Ramona. These are extremely beautiful and some could be well cut for gems, of rich yellow-green color. The crystals resemble in form those from Achmatovsk in the Ural, and are as perfect, brilliant, and transparent as those from Untersulzbachthal in the Austrian Tyrol. They are found in association with essonite garnet and quartz.

AGALMATOLITE.

H.= 2.5. G.= 2.81. Color grayish, grayish green, brownish and yellowish.

Agalmatolite, or Pagodite—a name given to some compact varieties of pinite (mica), pyrophyllite, and steatite—is like ordinary massive pinite in its amorphous texture and luster, but contains more silica. The name is from the Greek "agalma," an image.

The Chinese carve the soft stone into miniature pagodas, images, etc.

EL DORADO COUNTY.—A beautiful ornamental stone resembling the Chinese figure-stone is found two miles west of Greenwood, in a vein from six inches to a foot in thickness. S. M. B. 5300.

SAN LUIS OBISPO COUNTY.—A stone similar to this has been found in this county, but the location has been lost. S. M. B. 4060.

LEPIDOLITE.

Lithia Mica.

H.= 2.5–4. G.= 2.84–3. Luster pearly. Color rose-red, violet-gray, lilac, yellowish, grayish white, translucent. A complex silicate of alumina, lithia, potash, etc.; containing also fluorine and water. In rhombic crystals or plates, or compact granular massive.

The massive variety is used to some extent for ornaments, such as ash-trays, dishes, vases, paper-weights, etc.

INYO COUNTY.—Pink and white lepidolite, with azurite from the vein matter of Half-Dollar mine. S. M. B. 4262.

SAN DIEGO COUNTY.—The largest deposit of lepidolite in the United States is found at Pala. It dips with a pegmatite vein at an angle of 15 degrees. The average percentage of lithia is from 3 to 5. This vein also carries pink tourmaline or rubellite. Eleven hundred tons, worth $27,500, were shipped in 1901 from the Stewart mine. S. M. B. 1229. Another vein undeveloped has been discovered in the west side of Mount San Jacinto. S. M. B. 2773. Large bodies of lepidolite have also been uncovered at Mesa Grande and at Oak Grove, and the mineral is a constant associate of the gem-tourmaline and kunzite at all the mines where those species occur in that region. For further data, see Kunzite and Tourmaline.

The Pala are the greatest known deposits of lithia minerals known, and the product is used by the American Lithia Company of New York to obtain the lithia from it for lithia tablets and other medicinal uses.

The California State Mining Bureau exhibited in the Mines building, Louisiana Purchase Exposition, a magnificent pagoda, and walls of some exhibits were made out of wonderful specimens of this rubellited lepidolite. This beautiful lilac lepidolite and pink tourmaline combination is susceptible of a high polish and can be readily worked and turned, so that it ought to find a ready market for such objects as vases, dishes, pin-trays, paper-weights, etc.

ILL. No. 24. ENTRANCE TO LEPIDOLITE MINE AT PALA, SAN DIEGO COUNTY—
LEPIDOLITE AND RUBELLITE.

CHRYSOCOLLA.

H.= 2–4. G.= 2–2.2. Color mountain-green, bluish green, sky-blue, turquoise-blue, translucent to opaque. Hydrated silicate of copper.

This mineral when coated with or contained in quartz or chalcedony is occasionally cut as a gem.

INYO COUNTY.—In pseudomorphs after cerussite at the Aries mine, Cerro Gordo.

KERN COUNTY.—Beautiful crystals have been found near Randsburg that were mistaken for turquoise.

PLUMAS COUNTY.—Chrysocolla and malachite in alternating layers at Engels copper mine, Light Cañon. S. M. B. 5434.

SAN BERNARDINO COUNTY.--Valley Wells and New York Mountains.

SAN DIEGO COUNTY.—A specimen is in the State Mining Bureau, locality not given. No. 7187.

APATITE.

Asparagus Stone.

H. = 5.   G. = 3.17–3.23.   Luster vitreous.   Color sea-green, bluish green, violet blue; occasionally yellow, brown, gray or red.   A fluophosphate of lime, containing also some chlorine.

Opaque specimens have been found in the eastern end of the Kingstone range, in San Bernardino County.   It occurs also at the Dos Cabezas mine, near Jacumba, San Diego County.

FLUORITE.

Chlorophane. Pyro-emerald.

H. = 4.   G. = 3.18.   Luster vitreous.   Color range extensive, including white, yellow, green, violet, sky-blue, amethystine blue, brown, wine, yellow, rose-red, crimson and pink.   Fluorescent and phosphorescent when slightly heated.   Calcium, 51; fluorine, 48.9.

Fluorite, though too soft for continuous wear, is occasionally cut as a gem.   The massive varieties are worked up into paper-weights, vases, etc.   It is also known as chlorophane, pyro-emerald, fluorspar, Derbyshire spar, and Cabra stone.   Finely colored specimens are also known, according to color, as *false* ruby, topaz, emerald, amethyst, etc.

CONTRA COSTA COUNTY.—In white cubes at Mount Diablo.

MONO COUNTY.—At Ferris Cañon, in the Sweetwater Mountains. S. M. B. 14336.

SAN DIEGO COUNTY.—At Palomar Mountain, Oak Grove, a large piece of greenish fluorite has been found.

ANHYDRITE.

Vulpinite.

H. = 3.5.   G. = 2.9.   Luster vitreous and pearly.   Color white, grayish, bluish, reddish; also brick-red and blue.   Anhydrous sulphate of lime.   Lime, 41.2; sulphuric acid (sulphur trioxide), 58.8.

Anhydrite has also been called muriacite, tripe-stone, and anhydrite according to its structure—*muriacite* when crystallized in broad lamellæ; *anhydrite* when granular; and *tripe-stone* when composed of contorted plates.

Vulpinite is a scaly granular variety, and is the kind most used for ornamental purposes.

INYO COUNTY.—Panamint and Funeral range, Death Valley.

MONO COUNTY.—Mountains south of Mono Lake.

SAN BERNARDINO COUNTY.—Owl Mountains, near Owl Springs, Avawatz Mountains near to Amargosa River.

SAN DIEGO COUNTY.—Riverside range, Mesa Grande.

GYPSUM.

Alabaster. Satin Spar. Selenite.

H.=2. G.=2.3. Color white, gray, flesh-red, honey, ochre-yellow, blue, brown, and black. Sulphate of lime (hydrous). Lime, 32.6; sulphuric acid, 46.5; water, 20.9.

Selenite occurs either in distinct crystals or broad folia that are transparent.

Satin Spar is a fine fibrous variety having the pearly opalescence of moonstone and affording the cat's-eye ray when cut *en cabochon*. It is frequently worked up into beads, pins, and other ornaments, but is too soft to stand any wear.

Alabaster is a fine-grained, white, or delicately clouded variety. It is worked into carvings, statuettes, and other ornamental objects.

Gypsum forms extensive beds in connection with limestones and marlites; it is found also in crystalline rocks; about the fumaroles of volcanoes and in the desiccated lakes of the desert, and in the borax, soda, and niter fields, where it has been deposited on the evaporation of sea water and brines in which it exists in solution.

INYO COUNTY.—Common in many places.

LOS ANGELES COUNTY.—Soledad Cañon.

MONO COUNTY.—Abundant in desert portion.

RIVERSIDE COUNTY.—In desert portion, in large quantities and superior quality.

SAN BERNARDINO COUNTY.—Common in the dry lakes of the Mojave Desert.

SAN LUIS OBISPO COUNTY.—At Cholame.

SAN DIEGO COUNTY.—Abundant in Salton Desert.

SANTA BARBARA COUNTY.—Point Sal in large quantities and of superior quality, also at Rancho Casmalia.

COAL.

Jet. Lignite. Cannel. Anthracite. Brown Coal.

Jet.—The most important of the mineral coals used for ornamental purposes is jet, a compact, soft, light coal of a lustrous velvet-black color, susceptible of a high polish. It is a variety of lignite. A specimen from Gold Bluff, Humboldt County, is in the State Mining Bureau, No. 7321. It was taken from a carbonized tree in a coal vein.

Cannel Coal takes a good polish and is occasionally worked into inkstands, snuff-boxes, breast-pins, bracelets, etc.

Brown Coal.—In this the form and fiber of the original wood are preserved. It is made into paper-weights, charms, and trinkets. Specimens suitable for these purposes are often obtained from the coal mines in Amador, Alameda, Contra Costa, Monterey, Orange, and Riverside counties.

HEMATITE.

Including Chromite, Ilmenite, Limonite.

Hematite.—H.= 6.   G.= 4.9–5.3.   Iron, 70; oxygen, 30.   Color iron-black to steel-gray.

Limonite.—H.= 5–5.5.   G.= 3.6–4.0.   Color black, brown-black, gray. Iron, 60; oxygen, 25; water, 15.

Ilmenite or Menaccanite or Titanic Iron (oxide of titanium and iron).—H.= 5.6.   G.= 4.5–5.   Color black.   Oxygen, 31.6; titanium, 31.6; iron, 36.8.

Chromite.—H.= 5.5.   G.= 4.3–4.6.   Color black.   Iron oxide, 32.0; chromium oxide, 68.0.

The compact fibrous kinds of these irons are cut into beads, intaglios, charms, and other ornaments.

ALAMEDA COUNTY.—Hematite near Alameda; limonite in hills east of Alameda.

ALPINE COUNTY.—Hematite at Monitor.

AMADOR COUNTY.—Hematite in Ione Valley.

BUTTE COUNTY.—Hematite near Oroville.

CALAVERAS COUNTY. · Hematite and limonite at Campo Seco and San Andreas; limonite at Sheep Ranch and Big Trees.

DEL NORTE COUNTY.—Hematite at Kelsey tunnel.

EL DORADO COUNTY. – Hematite at Diamond Springs; limonite near Latrobe; beautiful compound crystals of ilmenite at Georgetown, from the placers; chromite in many places.

FRESNO COUNTY.—Ilmenite at Buchanan.

INYO COUNTY.—Hematite in Owens Valley.

MENDOCINO COUNTY.—Limonite. S. M. B. 7104.

NEVADA COUNTY.—Hematite at the Holden ledge in T. 15 N., R. 7 E.

PLACER COUNTY.—Hematite at Clipper Gap iron mine, and at Red Hill. Limonite in nodules at Forest Hill, resembling coprolites.

PLUMAS COUNTY.—Hematite at Crescent Mills, Mumford's Hill and Light Cañon.

RIVERSIDE COUNTY.—Hematite. S. M. B. 7107.

SAN BERNARDINO COUNTY.—Hematite, Iron Mountain and Bessemer Mountain, in the Mojave Desert.

SAN LUIS OBISPO COUNTY.—Hematite at Harrington iron mine, T. 31 S., R. 11 E.; also limonite.

SHASTA COUNTY.—Hematite and limonite at the Iron Mountain mine.

SIERRA COUNTY. – Limonite at Gold Lake.

SOLANO COUNTY.—Limonite on the shores of the bay, in nodules.

SONOMA COUNTY.—Hematite and limonite in quantity at Santa Rosa.

YUBA COUNTY.—Limonite on Bear River, near Wheatland.

GOTHITE.

H. = 5–5.5. G. = 4–4.4. Color yellowish, reddish, and blackish brown. Often blood-red by transmitted light. Oxide of iron, with 10 per cent of water.

This hydrated oxide of iron often occurs in acicular crystals penetrating limpid quartz. This form of inclusion is known as Onegite, and is frequently cut into gems.

MARIPOSA COUNTY.—At Burns' Creek, in quartz.

CASSITERITE.

Wood Tin. Toad's-eye Tin.

H. = 6.7. G. = 6.8–7.1. Luster adamantine. Color brown or black; to gray, white or yellow. Tin, 78.6; oxygen, 21.4.

Cassiterite, or tinstone, is used to a limited extent for ornamental purposes. The wood tin occurs in reniform or botryoidal shapes of concentric layers or bands resembling dark wood.

RIVERSIDE COUNTY.—Some of the crystallized tin ores from the Temescal district have been polished flat and resemble a dark polished wood.

BROOKITE.

Arkansite.

H. = 5.5–6.   G. = 3.87–4.08.   Titanium, 60.0; oxygen, 40.0.

Brookite does not polish readily, and this fact limits its use as a gem. Brookite includes the hair-brown, yellowish, ruby-red, transparent to translucent kinds having a metallic, adamantine luster. Arkansite includes the brilliant, iron-black, opaque kinds.

EL DORADO COUNTY.—Brookite implanted upon quartz crystals has been found at Placerville, by Mr. James Blakiston. The crystals of brookite are tabular, about two millimeters broad and one fourth of a millimeter in thickness. Their color is a rich reddish or yellowish brown. The ledge from which the mineral was obtained is a quartz ledge that is partly decomposed and partly compact. A decomposed quartz vein traverses the main vein for about 100 feet, and is filled in with reddish earth and sand. This decomposed material is full of quartz crystals from the size of a man's finger to those weighing 80 or 90 pounds, some perfectly clear and others with inclusions of green chlorite and of chalcedony.

AZURITE AND MALACHITE.

Blue Carbonate of Copper.   Green Carbonate of Copper.   False Emerald.

H. = 3.5–4.   G. = 3.8–4.0.   Azurite shows various shades of azure to Berlin blue. Malachite is a bright green. Malachite contains copper oxide, 71.9; carbon dioxide, 19.9; water, 8.2. Azurite contains copper oxide, 69.2; carbon dioxide, 25.6; water, 5.2.

Both are elegant minerals, common in copper mines, and when compact, especially malachite, are used for ornamental work, such as vases, table-tops, mantels. Malachite is most highly valued in Russia, where the greatest palaces and churches are embellished with it.

The uses of the two carbonates of copper as gems are limited by their softness. A favorite form is where the two minerals occur in alternating layers.

These minerals are found in nearly all the copper districts of the State; especially in the following localities:

CALAVERAS COUNTY.—At the Hedges mine, Copperopolis.

INYO COUNTY.—Coso Mountains.

KERN COUNTY.—At San Emigdio ranch with melaconite.

MONO COUNTY.—At Blind Springs. S. M. B. 4746.

SAN BERNARDINO COUNTY.—New York Mountains, Valley Wells.

SAN LUIS OBISPO COUNTY.—Santa Rosa Creek.

SANTA CLARA COUNTY.—With crystallized cinnabar in crystallized calcite at the Guadalupe mine.   S. M. B. 4929.

SHASTA COUNTY.—At Copper Hill.

TURQUOISE.

H.=6.   G.= 2.6–2.8.   Luster somewhat waxy.   Color sky-blue, bluish green and greenish gray. $P_2O_5 = 32.6$; $Al_2O_3 = 46.8$; $H_2O = 20.6$. A copper phosphate is also present, giving rise to the blue color.

FRESNO COUNTY.—At Taylor's ranch, Chowchilla River, several hexagonal bluish-green crystals, about one inch long, were found which were identified as turquoise by Dr. Gideon E. Moore and Prof. V. von Zepharovich.   The latter believed them to be pseudomorphous after apatite.*

SAN BERNARDINO COUNTY.—In the extreme northeastern part of this county there have been discovered old and abandoned mines of turquoise covering an area of many square miles.   Associated with these mines were found the relics of an early race; and it is supposed that this is the original source of much of the turquoise found in the hands of the Indians of the southwestern United States and Mexico.   The turquoise occurs in small veins and also in kidney-shaped masses about the size of a bean.   Much of it is of good quality.

The first published announcement of turquoise discoveries in this region was made through the writer in 1897, in his report to the U. S. Geological Survey.†   The locality was given as near Manvel.   Mr. T. C. Bassett had observed in this neighborhood a small hillock where the float rock was seamed and stained with blue.   On digging down a few feet, he found a vein of turquoise—a white talcose material inclosing nodules and small masses of the mineral, which at a depth of 20 feet showed fine gem color.   Two aboriginal stone hammers were met with, as usual at all the turquoise localities in the southwest, and from this circumstance the location was named the Stone Hammer mine.

The State Mining Bureau reported at about the same time that turquoise had been found in the desert region between Death Valley and Goff's Mining District, nearer the former, and that good samples were in the museum of the Bureau.

In the spring of 1898, many reports of extensive discoveries were

---

* Zeitsch. für Kryst. u. Min., Vol. X, p. 240.
† Min. Res. U. S., 1897, p. 504.

announced, and much attention was given by the press to the accounts of the region, both for the turquoise itself, and for the remarkable archæological remains associated with the ancient workings. The district was seen to cover quite a large area in northeastern San Bernardino County, near the Arizona and Nevada lines.

On the reports of prospectors reaching San Francisco as to a great group of ancient turquoise mines with cave dwellings, stone implements, and rocks covered with inscriptions, an exploring party was organized by the San Francisco "Call," and Mr. Gustav Eisen, of the California Academy of Sciences, became attached to it as archæological expert.* The party set out early in March, 1898, going first to Blake Station on the Santa Fé Railroad, thence north to Manvel, and onward some sixty miles, across the Ivanpah Sink, and up into the mountains to an altitude of over 6000 feet, through an exceedingly rugged country, to reach the region reported. The turquoise district, as described by Mr. Eisen and others of the party, occupies an area of 30 or 40 miles in extent, but the best mines are in a smaller section, about 15 miles long by 3 or 4 in width. The region is conspicuously volcanic in aspect, being largely covered with outflows of trap or basaltic rock reaching outward from a central group of extinct craters. These flows extend for many miles in all directions, and appear as long, low ridges, separated by valleys and cañons of the wildest character. Among these basaltic rocks and in the valleys are found smaller areas of low, rounded hills of decomposed sandstones and porphyries, traversed at times by ledges of harder crystalline rocks, quartzites, and schists. In the cañons and on the sides of these hills are the old turquoise mines, appearing as saucer-like pits, from 15 to 30 feet across and of half that depth, but generally much filled up with débris. They are scattered about everywhere. Around them the ground consists of disintegrated quartz rock, like sand or gravel, full of fragments and little nodules of turquoise. Whenever the quartzite ledges outcrop distinctly they show the blue veins of turquoise, sometimes in narrow seams, sometimes in nodules or in pockets. The mode of occurrence appears closely to resemble that at Turquoise Mountain, Arizona. A few prospectors have dug into the old, half-filled depressions and found stones of good color and quality, and ordinary ones may be picked up almost anywhere out of the decomposed quartz. Stone tools are abundant in the old workings, and the indications are plain that this locality was exploited on a great scale and probably for a long period, and must have been an important source of the turquoise used among the ancient Mexicans. From an archæological point of view this locality possesses remarkable interest. The cañon walls are full of caverns, now filled up to a

* See 20th Rept. U. S. Geol. Surv., Min. Res., 1898, pp. 582-584; and San Francisco "Call." March 18, 1898.

depth of several feet with apparently wind-blown sand and dust, but whose blackened roofs and rudely sculptured walls indicate that they were occupied for a long time by the people who worked the mines.   In the blown sand were found stone implements and pottery fragments of rude type, incised but not painted.   The openings to these caves are partially closed by roughly built walls composed of trap blocks piled upon one another with no attempt at fitting and no cement, but evidently made as a mere rude protection against weather and wild beasts. The tools, found partly in the caves and largely in the mine pits, are carefully wrought and polished from hard basalt or trap, chiefly hammers and adzes or axes, generally grooved for a handle and often of large size.   Some are beautifully perfect, others much worn and battered by use.

The most impressive feature, however, is the abundance of rock carvings in the whole region.   These are very varied, conspicuous, and peculiar, while elsewhere they are very rare.   Some are recognizable as "Aztec water signs," pointing the way to springs; but most of them are unlike any others known, and furnish a most interesting problem to American archæologists.   They are numbered by many thousands, carved in the hard basalt of the cliffs, or, more frequently, on large blocks of the same rock that have fallen and lie on the sides of the valleys. Some are combinations of lines, dots, and curves into various devices; others represent animals and men; a third and very peculiar type is that of the "shield figures," in which complex patterns of lines, circles, cross hatchings, etc., are inscribed within a shield-like outline perhaps 3 or 4 feet high.

One curious legend still exists among the neighboring Indians that is in no way improbable or inconsistent with the facts.   The story was told Mr. Eisen by "Indian Johnny," son of the Piute chief, Tecopah, who died recently at a great age, and who in turn had received it from his father.   Thousands of years ago, says the tale, this region was the home of the Desert Mojaves.   Among them suddenly appeared, from the west or south, a strange tribe searching for precious stones among the rocks, who made friends with the Mojaves, learned about these mines, and worked them and got great quantities of stones.   These people were unlike any other Indians, with lighter complexions and hair, very peaceable and industrious, and possessed of many curious arts. They made these rock carvings and taught the Mojaves the same things. This alarmed and excited the Piutes, who distrusted such strange novelties, and thought them some form of insanity or "bad medicine," and resolved on a war of extermination.   After a long and desperate conflict, most of the strangers and Mojaves were slain, since which time, perhaps a thousand years ago, the mines have been abandoned.   Mr. Eisen connects this account with the existence of a fair and reddish-haired tribe, the Mayos (not Mayas), in parts of Sinaloa and Sonora, some of whom may have reached these mines and carried on a turquoise trade with Mexico.

This region has since been opened at several points, and at least a dozen mines are now being worked by various parties, mostly with Eastern capital. The principal work is being done by the Himalaya and the Toltec mining companies. The turquoise obtained, when pure and of good color, is cut into fine gems; also the white and blue combination known as turquoise matrix, when small portions and veins of turquoise are distributed through the rock, and the whole is cut and polished as an ornamental stone. The paler varieties of turquoise are cut into beads, etc., long strings of which are sold. Most of the material produced is sent to New York. The yield in 1900 was estimated at a value of $20,000.

AMBER.

Succinite.

H.= 2–2.5. G.= 1.05–1.09. Brittle. Luster resinous and waxy. Transparent to opaque. Negatively electrified by friction. Burns readily with a rich yellow flame and aromatic odor. Pure succinite is not soluble in alcohol. Carbon, 78.94; hydrogen, 10.53; oxygen, 10.52. Color yellow, sometimes reddish, brownish, or whitish, often clouded, occasionally fluorescent, exhibiting a peculiar blue or green tinge.

Amber is a fossil resin of vegetable origin. Impure specimens have been found in several of the lignite coal veins of the State, but none of the true gem character. Its use is principally for beads, necklaces, etc.

CARBONATE OF LIME.

Pearl, Marble, Calcite, Aragonite, etc.

Carbonate of lime is most widely distributed in a variety of forms depending upon differences in origin, crystallization and structural conditions, presence of impurities, etc. With the exception of pearl and coral, the many kinds find a use more for decorative purposes than for personal adornment.

Pearls are concretions possessing a luster peculiar to themselves, found in the shells of certain mollusks. H.= 2.5–3.5. G.= 2.5–2.7. They may be of any shape and in some cases of considerable size. In color they range from an opaque white, through pink, yellow, salmon, fawn, red, purple, green, brown and black, or iridescent. Their beauty and value are dependent upon their color, texture or "skin," transparency or "water," luster and form. The most valuable are those that are spherical or pear-shaped, slightly transparent, free from specks or blemishes, and possessing the characteristic luster in the highest degree. The pearl-oysters of the Pacific and Indian oceans have yielded the bulk of the pearls of the world. The pearl has been highly prized through all ages, and has long been the emblem of purity, beauty, and nobility. The abalone, also known as the haliotis or earshell, secretes curious pearly masses, sometimes of fine luster and value.

Marbles consist essentially of carbonate of lime, with more or less carbonate of magnesia. They are fine to coarse granular in structure, and exhibit various colors, as white, yellow, red, green, blue, etc., being often clouded and giving a handsome effect when polished. Statuary marble must be pure white and fine-grained; architectural marble may be white or colored. *Cipolin* is white tinged; *Sienna* is yellow, veined or clouded with brownish green; *Mandelato* is light red with white spots; *Bardiglio* is gray with dark clouding; *Verde-Antique* (mixture of marble and serpentine) is clouded yellowish to bluish green; *Porter* or *Egyptian* is black, veined with yellow; *Lumachelle*, or fire marble, is a dark brown shell marble with fire-like internal reflections; *Madreporic* contains corals; *Ruin Marble* shows, when polished, figures resembling ruined castles, etc.; *Oölite* is made up of grains resembling fish roe; *Pisolite* is like oölite, but the concretions are larger; *Stalactites* are the pendent masses formed in caves; *Stalagmites* cover the floors of caves; *California onyx, Oriental onyx, onyx marble, Mexican onyx*, etc., have beautiful banded, mottled, or cloudy structures, often showing wide ranges of colors.

Calcite has a hardness of 3 and a specific gravity of 2.72.

Aragonite has a hardness of 3.5 and a specific gravity of 2.93.

Satin Spar is fibrous crystalline calcite or aragonite, showing a beautiful sheen when cut into ornamental objects such as paper-weights, beads, etc. The cat's-eye like effect is very pleasing. Used extensively for this use in England and the Russian Urals.

AMADOR COUNTY.—Stalactite and stalagmite in numerous caves; the same in Calaveras County and Shasta County.

COLUSA COUNTY.—A specimen of aragonite from this county is in the National Museum at Washington, D. C. It is a single cabochon cut, of brown color, and measures 27 by 14 by 7 millimeters. Catalogue number, a–597:84, 114. There is also a polished slab of the same color; catalogue number, b–840:48, 540.

INYO AND MONO COUNTIES.—These counties have large resources of marbles and onyx. Undeveloped oölite has also been discovered.

NAPA COUNTY.—Aragonite or California onyx, or "Zem-zem," is represented in the State Mining Bureau by Specimen No. 14768, from near Zem-Zem, exact locality unknown.

SAN BERNARDINO COUNTY.—Large quarries of choice Verde-antique are situated near Victorville, on the north side of the San Bernardino range. At Colton, both Egyptian and Bardiglio are quarried. These marbles are described in the "Mineral Resources of San Bernardino County." Specimens in the State Mining Bureau are 11350 and 11424.

SAN DIEGO COUNTY.—Aragonite at Los Peñasquitos Creek. S. M. B. 7320. Oölite and Madreporic marbles, Salton Desert and Carizzo Creek.

SAN LUIS OBISPO COUNTY.—Onyx marbles, aragonite or California onyx quarries are situated in Sec. 9, T. 32 S., R. 15 E. Beautiful specimens are on exhibition in the State Mining Bureau (2006), and also in the Memorial Museum, Golden Gate Park.

SISKIYOU COUNTY.—The Griffin onyx quarries are situated 6 miles south of Berryvale. Specimens in the State Mining Bureau are 7355 and 8969.

SOLANO COUNTY.—A number of polished slabs of aragonite are on exhibition from the Suisun quarries at the State Mining Bureau, 2261 (this is a fossil), 386, 556, 670, 5345, etc. Aragonite is also found in considerable quantities at Vacaville. S. M. B. 5345.

SONOMA COUNTY.—There is an aragonite quarry at Healdsburg.

The aragonites of this State are locally known as California onyx. Only the best known localities are mentioned above.

ORBICULAR DIORITE (NAPOLEONITE).

A mass of orbicular diorite is situated in San Diego County, in Sec. 15, T. 16 S., R. 1 E., about two miles west of Dehesa. In part this rock is largely made up of spherules often $2\frac{1}{2}$ inches in diameter, as shown in Illustration No. 4, page 15. The rock itself is very dark colored, and when cut and polished makes a handsome stone. A recent correspondent of the press describes the rock as several hundred feet thick, forming a blanket on the side of a small hill some 1500 feet above the bed of the Sweetwater River, thinning out toward the top.

The orbicular diorite is of various shades of gray to olive-green and black; the round nodules are of some two inches diameter, with a finely crystalline radiate and concentric structure. Other portions of the rock are marked like the grain of wood, in black and green wavy lines, at times expanding into the likeness of knots, and also in curious "bolts," parallel, curved, or crossing each other, in great variety.

This stone is a well-known variety, but rare, and when polished is very striking. It occurs in Corsica, and has hence been called Napoleonite. Fine slabs from several localities are in the American Museum

ILL. No. 25. ORBICULAR DIORITE MINE, DEHESA, SAN DIEGO COUNTY—NEAR VIEW OF OUTCROP.

ILL. No. 26. PEBBLE BEACH AT PESCADERO, SAN MATEO COUNTY.

of Natural History at New York, and are both elegant and peculiar. This Dehesa locality, from the specimens seen, ought to yield a very fine product for use in the ornamental arts. It was described by Prof. A. C. Lawson in a paper read before the Geological Society of America, December 19, 1901.*

CAT'S-EYE.

The term cat's-eye is applied to a number of minerals which, when cut *en cabochon*, exhibit a peculiar opalescence characterized by a line or ray of light across the stone, resembling the contracted pupil of the eye of a cat. Among the minerals which when fibrous or cut across the cleavage will show the cat's-eye ray are: beryl, chrysoberyl, especially the cymophane; corundum, crocidolite, dumortierite; quartz filled with acicular crystals or fibrous minerals, such as actinolite, byssolite, hornblende, etc.; hypersthene, enstatite, bronzite, aragonite, gypsum, labradorite, limestone, hematite, etc. Such gems may be opaque, translucent, or transparent, and of any color.

HUMBOLDT COUNTY.—Actinolite cat's-eye at Eureka.

SAN MATEO COUNTY.—Quartz cat's-eye at Pescadero Beach.

SAN DIEGO COUNTY.—Quartz cat's-eye at Point Loma, fine tourmaline cat's-eyes at Mesa Grande (see p. 60), and beryl cat's-eyes at Rincon.

OBSIDIAN.

Obsidian, a peculiar glass-like stone of volcanic origin, essentially feldspar in composition, is found along Pit River, where handsome specimens of the streaked variety known as marekanite or "mountain mahogany" are found; also in Owens Valley, where it occurs in red fragments, and also banded with alternate layers of black and brown. Similar observations have been made by earlier and later travelers, among whom was the late Hon. Caleb Lyon, who in 1860 found the Shasta Indians of California making arrowheads from obsidian as well as from the glass of a broken bottle. In a letter which was published by the American Ethnological Society, he describes the method of manufacture.† It was quite a favorite stone with the aborigines of the West and Southwest, where it is somewhat widely distributed, especially in the Yellowstone Park and in Mexico; in the latter the Aztecs used it for much remarkable work in knives, spearheads, and ornaments. Mr. W. H. Trenchard reports obsidian in considerable quantity in San Felipe Valley, northeast of Julian, San Diego County; and it may doubtless be found more or less at many points in the State.

---

*Science (N. S.), Vol. XV, 1902, p. 415.

†Bull. Am. Ethn. Society, Vol. I, p. 39, New York, 1861.

PEARLS.

The most important marine pearl-fishery on the American continent is that of Lower California, the central point being at La Paz. Here the true pearl-oysters, *Meleagrina margaritifera*, are found on the eastern shores of the Gulf of California from Cape San Lucas to the mouth of the Colorado River, taking in about 1500 miles of coast, including the gulf islands. They are also found from La Barra de Ocoz, which is the boundary line between the republics of Guatemala and Mexico, to Mazatlan, a distance of 2000 miles, making for the pearl fisheries a total extent of 3500 miles.

These fisheries have recently been confirmed to the Pearl Shell Company of San Francisco, by special franchise from the Mexican Government. The beds were first discovered some three centuries ago by Hernando Cortez when he crossed to the Pacific and discovered Lower California, and the name of California, derived from "calidus," hot, and "fornius," a hearth, it is believed, is due to this journey, having been given by Cortez, who found the heat intense when he first touched California soil. He took possession of the fisheries, and sent a number of fine pearls to the Queen of Spain, subsequently requiring all fishers to send to the Blessed Virgin one tenth of all they found, and one tenth to the King of Spain. After some intermittent work, the fisheries, about one hundred and fifty years ago, were again worked, with system and with great success, by one Juan Ossio, who took from them yearly from 300 to 500 pounds of pearls, actually packing them on mules and selling them by the bushel. The shells were all brought up by head divers, and pearls were taken from them so plentifully that they became of comparatively small value. This heavy drain had the effect of rapidly diminishing the supply, and it is only of late years that fishing has again been carried on systematically. At present numerous beds are known and worked, at Loreto, off Point Lorenzo, the island of Cerrabro, the harbors of Picheluigo, La Paz, and in fact the whole west coast of the Gulf of California from La Paz to above the island of Loreto, and in the east the island of Tiburon, and the land above and below that island. All these places have been famous for their pearls.

In 1860, in order to conduct pearl-gathering in a more scientific manner, the owner of the Mexican grants, Señor Navarro, procured from San Francisco, Cal., a number of expensive schooners, with surf-boats, professional divers, and costly apparatus. After several years' experience he found that his experts, with their expensive outfit, were no more successful than the naked Indian divers, while the exorbitant wages demanded by them so diminished his profits that he wisely went back to the primitive methods followed by his ancestors. At present those ship-owners who undertake the fisheries on a large scale use apparatus imported from France and England, by means of which each man is able to bring up daily 300 pearl-oysters. The men employed are powerful Mexicans, and every diver has five assistants. Four men work the air-pumps for the suited diver, and the fifth attends to the life-line, letting down the diver and hauling him up, as well as hoisting up the nets or baskets full of shells and lowering the empty ones. The pump-men are fed and housed, and receive $15 a month; the life-line man is similarly looked after, and receives $25 a month; the diver receives $45 a month, and one tenth of all he brings up, netting him as high as $500 a month, if he is fortunate. Connected with each fishing party is a schooner of from 60 to 200 tons burden, and two or three small boats. The men live on the schooner during the entire six months. In addition there are numerous divers who work independently, and who show wonderful skill and aptness in their work. Generally, with no other appliance than a heavy stone attached to the waist, they plunge naked to the bottom, select suitable bivalves, and gather them into a bag, remaining under water as long as two minutes. The shells containing the pearls vary in diameter from 2 to 8 inches, 6 inches being the average size. They are found on hard rocks or on sandstone at the bottom of the sea, usually in bunches, holding to the rocks by a fibrous beard (byssus), the circular opening being on top and the shells usually a little open. The oysters are vertical, not lying on the flat. Each diver has a knife, with which he cuts a bunch loose and places them in a basket or net by his side; this is hoisted up when full, an empty one descending at the same time. On rising to the surface, the fisher empties his bag into one of the waiting surf-boats, which crafts, under careful guard, deliver their loads to a well-armed schooner, the latter vessel running in shore at night to discharge the accumulated cargo. Occasionally, during all the time he is under water, a man may not send up a single shell containing a pearl; at other times there may be $10,000 worth in twenty shells. A very strict police system is necessary to prevent serious thefts, which, despite the utmost vigilance, are of daily occurrence. On land the cargo is turned over to keepers, and the mass is surrounded by guards, armed to the teeth. The shells are pried open with a flat knife, and the mussel is separated from each shell. A gristly substance attaches the body of the oyster to the shell, and covers about one fourth of its area, the remainder being occupied by the pearl-bearing membrane, a black, jelly-like coat, and, of course, a part of the living shell-fish. The shells are handed over to another man, while the opener takes the separated fish and examines the inside of the black membrane for the pearls he is in search of, and finally closes his fist over the fish to squeeze out any pearl which may be lodged in the interior, after which the pearls found are examined by experts, their value estimated, and a settlement made at once with the divers.

Usually their wages amount to twenty-five per cent of the total find, and they are paid by an allotment of the pearls taken during the day. On the outside the shells are covered with seaweed or other submarine growth, and look not unlike a Tam-o'-Shanter cap. All this growth is removed and the shells are cleansed and packed, finding a ready market in Liverpool, London, and Hamburg at prices of from 10 to 20 cents a pound for "mother-of-pearl." The profit from these fisheries is not as large as might be imagined, because the expenses are very heavy, and there is always involved a very considerable element of chance.

About 1863 a company was organized in New York City for the purpose of gathering pearls and pearl shells on the Pacific coast, and secured the use of a submarine boat, the peculiarities of which were that it carried a large supply of fresh air condensed within its walls and was provided with a means of purifying the air in the working chamber, thus dispensing with the necessity of communicating with the surface, as it furnished an atmosphere in which men could work for a whole day with perfect ease. The company procured a lease of property at the island of Tiburon, hoping, with their facilities, to secure unusual returns; for, with their submarine boat, they would have the advantage of exploring, locating, and working beds where divers could not go. Presumably their efforts were not successful, for the company soon went out of existence.

During the subsequent summer a new company obtained the concession for the Lower California pearl fisheries, and they decided that all the fisheries on the Gulf of California should in the future be worked by Chinamen.

For more than three hundred years these fisheries have been in the possession of private grants dating back to the days of the Conquest. The Mexican Government has in recent years annulled the old grants and leased the fisheries to the highest bidders. The house of Gonzales & Ruffo, having offices in La Paz and the City of Mexico, secured a concession for sixteen years permitting them to work the fisheries around the Espiritu Santo and La Paz islands, which are considered the best of the beds. The Government has recently granted to a single firm the exclusive right to raise the mother-of-pearl shells, and for the reproduction of such oysters the system used in the State of Maryland will be followed. The fisheries, which constitute one of the leading industries of Lower California, are now diminishing yearly, and, owing to the continued exploitation, many of the ship-owners find themselves losers at the end of the season.

Most of the pearls from this place are sent to market by way of San Francisco. A letter to the author from a leading fishing firm in 1892 contains the following: "The pearl fisheries average about 5,000 carats a year, which represent a value of $200,000, to which you must add about

800,000 pounds of pearl shells, representing a value of about $180,000. The cost amounts to about $100,000." During 1887 it is believed that more than $50,000 worth of pearls were found. The total product of the fisheries has amounted to as much as $250,000 in a single year, and the sale of the shells to as much more. From November, 1868, until September, 1869, $26,000 worth of pearls were purchased from this locality by one New York house. These were of various sizes, including four that weighed over 20 grains and one of 49 grains. In color, the pearls from this locality vary from pure white through gray and brown to black. The latter have become so fashionable in late years that their value has increased tenfold. One black pearl weighing 50 grains was valued at $8,000. A magnificent pear-shaped pearl of a less size was held at £7,000 in 1904. Black pearls and gray pearls, when fine, are among the most highly prized products of the sea.

ABALONE.

The Abalone (Haliotis or Earshell), the principal species of which are *Haliotis splendens* and *Haliotis rufescens* (called *ormer* in the Channel Islands, *fuh-yu* in China, *awabi* in Japan, and *abalone* in California), also secretes pearls. The nacreous portion of the shell itself is used for ornamental purposes, such as buttons, etc., and surface ornamentation in lacquer work, papier-maché, etc.

The fishing is conducted at low tide, the principal grounds on the coast being along the Catalina and Santa Rosa islands, in the Santa Barbara channel, and from Monterey to San Diego, although a large number are gathered in Halfmoon Bay and from the rocks that line the shore of Mendocino County. The earshells attach themselves to the rocks by means of their large muscular disk-shaped foot (so called), which acts like a sucker or exhaust cup. Just before the tide leaves them on the ebb, and just after it has reached them on the flow, the abalones keep their shells slightly raised above the surface of the rock with the feelers drawn in. Then the fisherman, with either a long, broad knife or a spade-like instrument—both are used—gives a quick lift to the sucker or foot, letting in the air. The suction is destroyed and the fish falls off, when it is seized and thrown into a boat or basket, before it can fasten itself afresh. If the fish lies below water, a sort of grappling iron is let down, and after the point is inserted under the shell a vigorous wrench pulls it away. All this has to be done quickly and quietly, for if the abalone closes down on the rock, it can not be drawn off, so great is its power of adhesion, and it will be broken into fragments before it releases its hold. When caught, the abalones are thrown on the beach, and the fish is pulled from the shell with a flat, sharp stick, and stripped of its curtain, boiled, salted, and strung on long rods to dry in the air. This process is very disagreeable, and that of stripping and cleaning so offensive that none but Chinese will undertake it. The abalones must be as hard as sole-leather when properly dried, and they are then packed in sacks, and sent to China. The price of the meat is from 5 to 8 cents a pound in San Francisco, or from 7 to 10 cents a pound in China. When cooked, it is cut into strips and boiled, the taste being similar to that of the clam, but with a more meat-like consistency.

The trade in this dried meat is considerable. In 1866 there were exported from San Francisco by steamer 1697 sacks, valued at $14,440, and in 1867 the exports had risen to 3713 sacks, valued at $33,090. At present there are exported upwards of 200 tons a year, which at $175 a ton would amount to $35,000. At San Diego, Cal., the dried meat is quoted at $110 a ton. The shells vary from almost microscopic size to 8 or 10 inches in diameter. Before they were found to be of any marketable value they were thrown away. One heap a little south of San Diego, containing over a hundred tons of shells, from exposure to the rain and the sun was converted into lime on the outside, but this was broken into and many fine shells were found.

The shell in its natural state is no more attractive than that of the oyster; it is rough on the outside, looking much like a piece of dried brown clay, and is frequently covered with a growth of barnacles, seaweeds, etc. Commercially there are five varieties, the green, the black, the red, the pink, and the mottled; but considering them from an ornamental standpoint, the shells may be grouped under three heads, red, black, and green, so called, of course, from their prevalent color. The black, which is the smallest and least valuable, is found from Monterey down to the Gulf of California; the red, which is next in value, but the largest in size, is found from Mendocino to Monterey; while the green comes from below San Diego. The black seldom exceeds 6 inches in diameter, the green rarely goes beyond 9, while the red runs as high as 12 or 14 inches. The black is not beautiful on the outside, even when cleansed of lime and marine parasites, but inside there lies a small patch of the most brilliant opalescent tints, and this is sawn out, and made into brooches and lockets. The red is of a general mother-of-pearl appearance, with stripes and mottles of a rich burnt umber. The green, both within and without, is full of fire and color, some interiors being quite as vivid and of much the same prevailing color as a peacock's neck. This variety is principally used for jewelry, and is worked into every kind of ornament, from a table-top, inlaid with representations of flowers and butterflies, to the smaller varieties of jewelry. The Pueblo, Zuñi, and Navajo Indians, and all the Indians of the Pacific coast as far north as Alaska, have made it into charms and have used it for ornamentation for ages. It has been used as an applied decoration on silver objects, and examples were exhibited at the World's Fair held at Paris in 1889.

The first adaptation of the abalone shell to ornamental purposes was made by an English worker in mother-of-pearl who went to San Francisco more than twenty years ago. He saw the possibilities of the wonderful, brilliant shell, and began a business which now requires the services of more than ten men. The little trifles made of this shell are considered by the Eastern visitor and the European tourist as distinctively Californian as a piece of big-tree bark. The incrustations were formerly removed by soaking the shells in a bath of muriatic acid, but it was found that this process injured the texture, and they are now cleaned and polished by friction lathes. Twenty years ago abalone shells were considered so worthless that freight steamers would not transport a bag of them without advance payment for the freight. Now they are worth $150 to $175 a ton in New York and Liverpool. The shells are shipped first to San Francisco, where they are assorted and the damaged ones thrown aside, about three tons of merchantable shells being procured from five tons of material as it comes from the abalone-hunters. These shells are quoted (1889) in San Diego at $20, $25, and $35 a ton, according to quality, and in consequence of such low prices the trade is comparatively dull. The output of shells during 1888 was estimated at 300 tons. The amount of shells made into jewelry in San Francisco is very small compared with that consumed by the button-makers of France, England, Germany, and New York. Orders for abalone shells are constantly received from these places, and there are times when the export reaches as high a figure as 100 tons a week. The collector of customs at San Francisco furnishes the information that for the fiscal year 1887–88 the export of abalone shell amounted to $185,414, which together with $35,000, the value of the dried meat annually exported, makes this quite an important industry.

These shells secrete very curious pearly masses, sometimes of fine luster, and choice enough to deserve a place among pearls. A pearl measuring 2 inches in length, and from a quarter to a half inch in width, has been found. A necklace made in California from the finest specimens was valued at over $2000. A pearl over half an inch long and of good color cost $30, and was used as the body of a jeweled fly. The abalone pearls from the coasts of Korea and Japan are often very beautiful. In a lot of about one hundred shells only five were found bearing pearls—two with three pearls each, two with two pearls each, and one with a single pearl.

GEM MINES IN CALIFORNIA.

Besides the references made to a number of gem mines in the body of this Bulletin, the following more specific data are here presented with reference to some of the more important ones. For the purpose of obtaining this information, a special inquiry was undertaken at the close of the last year, 1904, in behalf of the State Mining Bureau; and representatives of the Bureau visited a number of the mines and collected valuable data, which are herewith presented.

It is well to remember the fact that there is already more actual mining for gems done in the State of California than in any other State or Territory of the Union, while the indications are that there will be many more gem mines discovered in southern California as remote districts are opened and old ones more fully explored.

The following data are grouped (1) geographically, and to some extent also (2) in the order of discovery—beginning in Riverside County, and proceeding southward and southwestward, in San Diego County, by Pala, Mesa Grande, and Ramona, to the Mexican line at Jacumba.

COAHUILA DISTRICT,

in Riverside County. These are the most northern occurrences of gem-tourmaline, and the earliest discoveries were made here.

Fano Kunzite-Tourmaline Mining Company.—This mine consists of four claims, about 3 miles north of Coahuila Indian Reservation, Riverside County, and was located in 1902 by Bert Simmons. The nearest postoffice is Hemet, Riverside County. After some surface work had been done, a tunnel was started 300 feet from the summit of the hill, to cross the ledge, but by a mistake in calculation the ledge proper was crossed about 20 feet from the surface. The parties then continued their operations until, at a depth of 176 feet, solid blue granite was reached. The tunnel was then abandoned, and from that time work has been confined to the surface.

The ledge is about 5 feet in width, with a northwesterly and southeasterly strike, and a dip to the southwest of about 17 degrees. The pegmatite is finely crystallized, and resembles that of the other tourmaline and kunzite mines in southern California.

Three men are at work at present, and operations will be continued indefinitely. The output so far has been 25 pounds of kunzite, white;

1 pound of kunzite, pink; and 25 pounds of all classes of tourmaline, mostly blue and green; about 250 pounds of beryl have also been taken out, but only about five per cent of it available for gem purposes. Two hundred pounds of very fine quartz crystals also have been sold, and about a ton of lepidolite and 30 or 40 pounds of amblygonite; also splendid flake mica large enough for commercial purposes has been discovered.

ILL. No. 27. COAHUILA MOUNTAIN (GENERAL VIEW), RIVERSIDE COUNTY. BERYL, KUNZITE, AND GEM-TOURMALINE.

There is a spring near the property on land rented by the owners of the mine; also plenty of oak timber for mining purposes. Considerable money has been expended here without much result, but for the work actually done on gem pockets, this mine has been a splendid producer.

Columbia Gem Mine.—This, the oldest tourmaline mine in the State, is situated at Coahuila, Riverside County, and owned by Messrs. H. C. Gordon, P. E. Johnson, J. C. Connell, and William Dyche, of San

Diego; it is about half a mile northwest of the road leading from Coahuila to the Hemet reservoir, and near the summit of the divide crossed by this road. Nothing has been done on this mine, except assessment work, for over five years, but it was the first tourmaline mine discovered in southern California, and has produced a great many beautiful gems. The pockets, however, seem to have been worked out, and nothing important has been found recently. The ledges of pegmatite are very fine granite, and both sides of the pocket material seem to be of the same character, thus differing from any other mine yet found in the gem districts of California. There is no water or timber available, and it is altogether a desolate region. The altitude is about 5000 feet.

ILL. No. 28. FANO (SIMMONS) MINE, COAHUILA, RIVERSIDE COUNTY—VIEW OF RIDGE, LOOKING NORTH. GEM-TOURMALINE, BERYL, KUNZITE.

Passing southward from the Coahuila region, into San Diego County, the locality next described lies by itself, about half way to the great Mesa Grande-Pala line of mines. Although not yet an important producer, the occurrence is very interesting, as suggesting other possible localities yet to be discovered in the intervening area.

Gem Mine No. 1.—Owned by Mr. Bert Simmons, of Oak Grove, and Mr. Charles Gordon, of San Diego. Practically no work has been done on this mine since its location in June, 1903. Its altitude is higher than any other gem mine in San Diego County, being 5100 feet above sea level, and about one mile east of the summit of Aguanga Mountain. The average width of the vein, as far as could be seen, was 4 feet, but it is badly broken, and upon examination showed that both foot and hanging walls were of very hard blue diorite. Great pressure has apparently crushed the ledge, and the pocket layer is found on the top, between the diorite and the pegmatite, and presents fine, broken crystallizations of orthoclase and albite, in which a red clay is mixed. The tourmaline crystals show much indication of dynamic action, being badly broken and twisted, but afford nodules of beautiful coloring— deep blues, reds, and an almost emerald-green predominating. The mine is located on the top of the divide or watershed between San Luis Rey River and the Temecula Cañon. So little work has been done that it seems better to reserve any report as to the quantity and quality until more is ascertained. Parties are at work at present on the mine.

PALA DISTRICT.

As elsewhere described in this Bulletin, the mines near Pala are located on three hills or ridges, the western being properly called Pala Mountain, on which are the great lepidolite, or Alvarado, mine, and the Stewart mine, next described, which yields some gem material. The other mountains, Pala Chief and Heriart, which are apparently foothills or spurs of Agua Tibia Mountain, are those yielding gem-spodumene as well as tourmaline. Some 18 miles to the southwest, but probably belonging to the same range of hills, lie the great tourmaline mines of Mesa Grande. These will be given in the order stated.

Stewart Mine, Pala Mountain.—This mine, said to have first been discovered by an Indian deer-hunter named Vensuelada, in the early days of California history, was first worked by a miner named Henry Magee, who located the claim as a quicksilver mine, mistaking the pink tourmaline for cinnabar, but upon analysis he abandoned his prospect. Next it was located as a rock-claim by Don Tomas Alvarado, a Mexican land-owner in that locality, who believed that the beautiful bluish, pinkish, and gray minerals studded with transparent pink crystals were a peculiar variety of marble. Several years later a German scientist, who was familiar with lithia mines in Europe, saw a specimen of Pala lepidolite in a mineral collection in New York. Obtaining a piece, he made an analysis and found that this ore was as rich in lithia as any found in the world. From this time forward, gradual development under many ownerships has proved that great deposits of lithia-bearing ores exist in the pegmatites of the Pala district, the largest and most valuable being the Stewart and Alvarado mines.

In examining the workings and surface of the Stewart mine, owned by the American Lithia Company, of New York, numerous indications of gem-minerals were met with, especially in the lower workings. As in the Alvarado mine, the lepidolite is generally studded with small, fan-shaped crystallizations of rubellite (pink tourmaline), with occasional crystals of bluish or greenish tourmaline, but not of gem quality. Near the surface the tourmalines are small and perfectly crystallized, but are more or less fractured, opaque, and unfit for jeweler's use. In the deeper workings and in the extreme western tunnels, however, pink tourmalines from one-half to one inch in diameter are found in columnar groups, all more or less altered, and of not over three (3) in hardness, associated with quartz, orthoclase, gray lepidolite,

ILL. No. 29. STEWART LITHIA MINE, PALA MOUNTAIN, SAN DIEGO COUNTY—WEST END OF TUNNEL, LOOKING NORTHEAST.

and amblygonite. Triplite and triphylite are also associated minerals. Large crystallizations of what appears to be an altered spodumene were observed, penetrating the quartz.

On the surface, small green tourmalines were found in the pegmatite, generally more or less flattened between the cleavage planes of muscovite mica.

Several years ago a pocket containing about a quart of small tourmaline crystals was found in coarse pegmatite, 60 feet south of the present tunnel of the Stewart mine. Some of these crystals were cut into very good gems, but no further work at that spot has been done.

Pala Chief Mine.—This mine was located in May, 1903, by Mr. John Giddens, Pedro Peiletch, Bernardo Heriart, and Frank A. Salmons. The main workings are at an altitude of 1220 feet (aneroid). The work consists of open cuts 250 feet wide, extending to a depth of from 10 to 30 feet horizontally on the vein, and at the deepest working the ledge is 21 feet in width vertically. A tunnel 45 feet long was run to encounter the vein up to about 20 feet depth, but it was found that the ledge was a blanket vein, and nothing was discovered in that place. But in the upper or surface workings the hanging and foot walls were both found to be of bluish and grayish decomposed diorite. The upper

ILL. No. 30. PALA CHIEF MINE, PALA, SAN DIEGO COUNTY—EXTENT AND CHARACTER OF WORK DONE IN DEVELOPING KUNZITE.

part of the vein consists of 3 feet of white, finely crystallized pegmatite. Beneath this the crystallizations become coarser and more granitoid. The third layer was composed partly of finely crystallized albite and orthoclase, upon the lower edge of which, and extending to the pockets, was a layer of lithia-bearing micas. In the interior of the pockets, which are generally 8 to 10 inches wide, pinkish and white talc was found, in which occurred numerous large and perfect quartz crystals with pink and white spodumene. As in most of the mines of southern California, the lower half of the ledge, below the pocket line, is a very finely crystallized granite without mica, with small crystals of essonite garnet. The above-described characteristics of the ledge are general throughout the mine.   The minerals noted were spodumene, pink, lavender, and white; tourmaline, blue, green, and red; orthoclase, albite, graphic granite; lepidolite, pink, green and lavender; muscovite, quartz crystals, steatite, and other clays.

The products so far noted are tourmaline, kunzite, and quartz crystals. Giant powder was used entirely, and it was found to be the only explosive that was satisfactory.   Two men have been working nearly all the time; but during the last six months very little of the precious stones rewarded their labors.   There is no water or timber on the property, and the nearest water is about one mile away.

ILL. No. 31. PALA CHIEF MINE, SAN DIEGO COUNTY—POCKET LINE AT THE POINT WHERE THE LARGEST POCKET OF KUNZITE CRYSTALS OCCURRED.

The section and township in which the mine is located were not available, but it lies east from Pala, at a distance of 3 miles, and the workings can be seen from the town of Pala, which is the nearest base of supplies.

Tourmaline Queen Mine.—This mine, owned by Mr. Frank Salmons, John Giddens, Pedro Peiletch, and Bernardo Heriart, is situated near the summit on the northeast slope of Pala Chief Mountain, at an altitude of 1450 feet.   It is about 3½ miles north by a little east from Pala, San Diego County.   The section and quarter were not obtainable.   The mine was located as a quartz claim by the above-named parties, in

March, 1903.   The vein is about 14 feet wide, and dips to the southwest 15 degrees.

Very little has been done on the property, but scalping work in the nature of an open cut 60 feet wide, and entering the vein to a depth of about 10 feet, produced in weight approximately 80 pounds of gem-tourmaline crystals.   The colors are yellow, green of several different shades, light pink, ruby-red, and black.   In examining the ledge, 18 inches lying between the diorite hanging wall and the coarse pegmatite appears to be an infiltration of decomposed feldspar, gradually altering to pegmatite.   Below this are about 3 feet of coarse, granular pegmatite (or granite), consisting of crystallized quartz, feldspar, and muscovite mica, with impurities of black tourmaline in fan-shaped crystallizations, and essonite garnets (microscopic), with occasional crystals of biotite mica and hornblende.   Below this again, and gradually altering from the above, are masses of graphic granite, incrusted at the lower edge with albite, in which the gem-tourmaline seems to have a root or extremity.   Between the albite and the line-rock (or granite) are large pockets filled with rose- and lavender-colored muscovite, and decomposed spars in the nature of a whitish or pink clay; in these pockets the gems are found, broken in many instances, and more or less altered.   Many crystals were observed with an exterior of opaque green, while the interior was a rich pink or ruby-red, affording beautiful gems.

The ledge has been prospected for about 250 feet, and shows gem indications wherever it has been opened.   The hanging wall is a coarse, greenish and grayish diorite, which is the general formation of the entire belt.   The foot wall is the same, though showing more alteration.   Both Giant and Judson powders have been used, although from the hardness and toughness of the rock, the former was found to be the best.

After the pocket material has been extracted, screens are used, by which the dirt and fine, worthless stuff are eliminated.   The matter left in the screens is then examined for gems, and afterwards washed. Two of the owners have performed all the work so far accomplished, and no other men have been employed.   Active operations will again be resumed, but nothing is being done at present.   The same parties have filed on a spring 350 feet northeast of the present workings, and abundant water for mining and domestic purposes has been developed.

The minerals noted in above claim are: tourmaline, albite, orthoclase, muscovite, lepidolite, kaolin, talcose clays, essonite garnets, hornblende, and indications of epidote.

The lower part of the ledge is composed of a fine, granular mica-less granite, of a gray color, banded at intervals of from 3 to 6 inches with minute essonite garnets, whence the name line-rock.   As is usually the case in all ledges of pegmatite bearing precious stones in this region, this lower layer of the ledge has approximately the same width as that of the formation from the pocket layer or center to the top, and lies directly in contact with the diorite foot wall.

Tourmaline King Mine.—This mine is owned by F. B. Schuyler, Mrs. F. B. Schuyler, D. G. Harrington, and Mrs. H. E. Harrington, of Oceanside, Cal., and is situated on the north slope of Pala Chief Mountain, about three hundred yards from the summit, at an altitude of 1540 feet. The mine was located in March, 1903, by the above-named parties, but very little work has been done, rendering it practically impossible to make a conclusive report. The mine is 4 miles directly north of Pala, and is the last mine so far discovered at the western extremity of the Pala mineral belt.

The vein dips to the southwest at an angle of 16½ degrees. It presents an average breadth of 7 feet, and is essentially coarse pegmatite, but shows evidence of crushing and is badly broken in many places. The hanging wall is a coarse gray diorite, and at the place where the work has been done lies over about 15 inches of coarse broken feldspar and lepidolite mica. It is in this stratum that the gems appear, which is contrary to the general pocket formation of the Pala district. Tourmaline was the only gem-stone noted, and occurred in pencils, disseminated through this altered mass of decomposed spar, and apparently out of place. Concretions of albite, coated with beautiful purple muscovite, were found loose in the soil. Some quartz crystals and essonite garnets, badly shattered, were also seen in the float. The ledge at this place was too badly broken to note the exact character of the pegmatite, and the "line-rock," or lower stratum, had not been uncovered, so that its character could not be determined. No work has been done on the property for several months, and nothing satisfactory could be learned as to when work would be resumed. About ten pounds of crystals were secured in a cut 12 feet wide, and barely scalping off the top layer of earth.

Naylor-Vanderburg Mine.—This mine, also situated near Pala, is owned by Fred M. Sickler and M. M. Sickler; altitude, 1400 feet, on the eastern slope of Mount Heriart. The location was made by Mr. Sickler in February, 1903, soon after he had discovered that the pink and white crystals which he had found on the mountainside were not tourmalines, as they had been called, or any stone known to local mineralogists. After much trouble and expense, Mr. Sickler considered the stone of uncertain value, but continued his investigations and at length sent a piece to the writer at New York, who determined it as spodumene, and after whom it was named kunzite, by Prof. Charles Baskerville, of North Carolina, as a new gem-stone—the first occurrence of transparent pink or lavender spodumene in the world.

9—MB

The ledge at the point examined was 16 feet in width, but badly broken. At this place an open cut entering the vein to a depth of 22 feet and about 30 feet in width, has produced approximately five pounds of perfect gem-stone; although several pieces have been found in adjacent workings, this seems to be the best part of the ledge.

In examination of the mine, the hanging wall is gray orbicular diorite. Between this and the ledge itself, an 18-inch layer of decomposed feldspar and clay was found as a gouge. About 7 feet of coarse granitic pegmatite forms the upper part of the ledge, altering into decomposed layers of albite and orthoclase. In this latter are small pockets, seldom

ILL. No. 32. NAYLOR-VANDERBURG MINE, HERIART MOUNTAIN, SAN DIEGO COUNTY—VIEW LOOKING SOUTHWEST.

larger than a man's hand, in which one or two crystals of kunzite will be found, completely covered with yellow, pink, or white clay. No metallic stains are found in the upper part of the ledge, but the lower beds of granitic rock, carrying interlineations of garnet, are in many places stained with manganese, and show large crystallizations of triplite, from which it is evident the kunzite receives its coloring.

The vein has a dip of 10 degrees to the west, and extends the full length of the location, 1500 feet, joining the Caterina mine on the south.

The minerals noticed are: muscovite, pink, green, and lavender, in

ILL. No. 33. NAYLOR-VANDERBURG MINE, SHOWING WORKINGS.

ILL. No. 34. NAYLOR-VANDERBURG MINE, HERIART MOUNTAIN, SAN DIEGO COUNTY.
"NAYLOR ROCK," SHOWING PEGMATITE ABOVE, ZONE OF POCKETS
AND BANDED "LINE-ROCK" BELOW.

very large scales; montmorillonite and steatite talcs; pink, green, and white spodumene; and black tourmaline, but no gems of that stone. Albite and orthoclase, with some potash feldspars, are the mother of crystallization. It has been reported that spinel has also been discovered in this mine; associated with it were deep-colored green beryl and columbite.

There is no water or timber on the property. This mining claim is embraced within the boundaries of the Pala Indian Reservation, but was located before the reservation was declared. The output of the

ILL. No. 35. MOUNT HERIART. (Taken from south, one mile.)
Cross (×) shows Naylor-Vanderburg mine. Square (□) shows Caterina mine.

mine since the beginning of work has been about ten pounds of gem-kunzite, no other minerals having been disposed of. Some pink and green beryls were noticed, but nothing has been developed in that line.

Other claims and openings on Mount Heriart are enumerated in the body of this Bulletin, this one being thus far the most important. The following are among the principal of these openings:—

Heriart Claim, owned by F. M. and M. M. Sickler. A tunnel has been run a distance of 40 feet. A pegmatite lithia-bearing ledge was encountered, from 1½ to 4 feet in width. The ledge occurs in a granite dike, which in turn traverses the diorite. The granite dike is about 100 feet in thickness, and can be traced for over 2000 feet. The lepidolite occurs in white and lilac colors, and is often full of radiated tourmalines, both pink and green. Large amounts of muscovite are often encased in the lepidolite. Amblygonite is also found. The tourmalines are of various shades of green, and some blue and pink crystals occur, but as yet have not been found in large quantities. Crystals of albite and orthoclase occur in the pockets. About two tons of lepidolite have been extracted from the tunnel.

San Pedro Claim.—On the San Pedro claim, owned by Pedro Peiletch and Bernardo Heriart, the Naylor-Vanderburg ledge has been cut. Some kunzite, tourmaline, and beryl have been found, besides lepidolite. This ledge is exposed for nearly a mile in length. At four different points it has been cut, exposing kunzite and lepidolite.

Caterina Mine, owned by Bernardo Heriart and M. M. Sickler. A cut has been made 40 feet in length and 30 feet in width, exposing a ledge of lepidolite 2½ feet in thickness. About six tons of lepidolite have been extracted from this cut. Some kunzite and spodumene were found, the greater part of which was float. Other gems found were pink beryl and a few tourmalines.

MESA GRANDE.

The Mesa Grande mines are situated on the hill or mountain of that name, and are the most southern of the gem-tourmaline localities in the region. The ridges stretch along northwestward to the Pala and Agua Tibia mountains, already described; to the west is another locality for tourmalines at Vista, and northward are, first, the Oak Grove location, and farther on those near Coahuila.

Several mines have been opened on the Mesa Grande, the Himalaya Mining Company occupying the west side of the ridge, and the San Diego Tourmaline Company the east side. The latter is working a property opened by Mr. Gail Lewis, at the time of the first discoveries on this mountain; he had but small success with it at first, but persevered, and reached a fine pocket of gem material just before his option expired. The mine has been developed more elaborately than any other, and carried much deeper. Fine gem-tourmalines are taken out here from a depth of 200 feet—the greatest depth at which these gems are obtained anywhere in the world.

Himalaya Mine.—This mine, owned by the Himalaya Mining Company, of New York, is situated in the E. ½ of Sec. 17, T. 11 S., R. 2 E., S. B. M., at an altitude of 3800 feet. The property is about 4½ miles northwest of the Mesa Grande store, and on the watershed between San

Luis Rey River and Mesa Grande Creek. For many years it has been known that beautiful colored stones existed on this ridge, but after repeated failures and with no determination of quality and value, the people of the locality gave up the property as worthless. At length an agent of Mr. Tannenbaum, Heighway by name, found the locality and recognized the stones as tourmalines. This led to developments by the Himalaya Mining Company, and the present output is the result.

During 1904 about six tons of rough tourmaline were shipped to the company's lapidary in New York; of this amount, 300 or 400 pounds were fine nodules and pencils of the very highest grade.

Surface or bench digging has been followed exclusively, although a

ILL. No. 36. HIMALAYA TOURMALINE MINE, MESA GRANDE, SAN DIEGO COUNTY.

tunnel is being run to tap the ledges at the 150-foot level. Both hanging and foot walls are of hard blue diorite, and the ledge is of fine crystallized pegmatite not over 18 inches in width, and dipping from 26 to 33 degrees southwest.

In working this ledge, pay material has been in sight continuously, and at no time has a barren piece of ground been-encountered.

The upper pegmatite is usually stained with lithia and manganese, and large masses of lepidolite are associated with tourmalines. The pockets are large and filled with talc and hydrous micas, in which the gem crystals occur embedded, many showing peculiar etchings. The ledge has been uncovered for about 700 feet, and to an average depth of 15 feet. These open cuts, however, are proving dangerous and will have to be abandoned as soon as the rainy season has soaked the walls on either side.

Among the minerals noted were orthoclase, albite, lepidolite, amblygonite, small clear pieces of spodumene, muscovite, tourmaline (black, green, blue, deep red, and rose), pink and aquamarine beryl, hornblende and epidotic rocks, spessartite and essonite garnet, large and very transparent quartz crystals, talc and hydrous micas, and a dark brownish transparent crystal, very dense (specific gravity, 10), and a hardness of 5½, which has not yet been determined. This mineral is very rare, and only a few pieces have been found.

Wood, water, and all natural advantages are of the best; and a good dwelling-house, barn, tool-houses, and blacksmith shop, as well as a windmill with water piped to all, constitute the improvements.

From four to ten men are constantly employed about the mine. The gross receipts for 1904 are estimated at $30,000.

San Diego Tourmaline Mining Company.—The mines are situated in the E. ½ of Sec. 17, T. 11 S., R. 1 E., S. B. M., and about 4 miles northwest of the Mesa Grande postoffice. Considerable work has been done on this property since 1901, perhaps more development work than on any other gem mine in southern California. In the first place, a tunnel 120 feet long was run, tapping the ledge at 64 feet. From this, drifts were run about 150 feet in either direction, and the ledge matter was stoped to the surface. Tourmalines in paying quantities were extracted, and from this output the San Diego Tourmaline Mining Company was organized. Later a tunnel was run 286 feet in length, tapping the ledges at from 145 to 170 feet, and drifts on two ledges which were struck from 20 to 30 feet. The ledge matter is a fine-grained pegmatite, showing on both top and bottom black tourmalines in fan-shaped crystallizations. Near the center, at intervals, pockets occur in which fine gem-tourmalines are found, but not as rich as in the adjoining claim, which is the property of the Himalaya Mining Company.

This company has employed from three to seven men continuously. They have a lapidary of their own in San Diego, where most of their product is cut.

Wood, water, and all facilities are at hand. Giant powder has been used exclusively, and has not resulted in the breaking or destroying of any crystals. The ledges are over 18 inches in width, and are generally of a character which would not be prospected, looking barren and worthless, but the locality seems to be highly mineralized and many ledges show gem crystals.

Other mines are being opened in the vicinity, and probably during 1905 there will be a great development in this particular section.

Esmeralda Mine, Mesa Grande.—The Esmeralda mine, owned by J. D. Stone and H. E. Dougherty, both of Mesa Grande, Cal., is situated about 5 miles northwest of the Mesa Grande store, and 1¼ miles west of the Himalaya mine, and on the eastern slope of the Temescal Valley, in the S. E. ¼ of the S. E. ¼ of Sec. 13, T. 11 S., R. 1 E., S. B. M. The mine was discovered May 7, 1904, by Mr. Dougherty, and was acquired by location as a quartz ledge. The altitude is 3470 feet.

The course of the ledge is northeast and southwest; but where the work is being done a spur running southwest and northeast at right angles with the main ledge has produced all the gems yet found. The

ILL. No. 37. ESMERALDA MINE, MESA GRANDE—TUNNEL LOOKING EAST.
GEM-TOURMALINE.

ledge dips to the southwest at an angle of 26 degrees and is about 10 feet in width at the point opened. The claim embraces one large ledge and numerous stringers, showing gem indications. The work at present performed consists of two open cuts crossing the vein and exposing it to a depth of 7½ feet; a tunnel 60 feet below the surface workings tapped the ledge at 28 feet; but no further work has been done in the tunnel, and no gems were found in the formation at that place.

Both hanging and foot walls are composed of a coarse, crystallized hornblendic diorite of a rich grass-green color, resembling a serpentine. The ledge itself is pegmatite, and is faulted in several places by volcanic action. The pegmatite is of the coarse granitic type met with in nearly all the gem mines in the southern belt. The pockets are quite large, and contain quartz crystals, orthoclase, and albite in beautiful transparent crystallizations. Lepidolite in pieces weighing from 50 to 300 pounds also occurs in conjunction with the pocket material. Tourmaline is the only perfect gem found, and occurs in pink, bright red, azure blue, aquamarine blue, and a peculiar shade of green blue, which cuts to a stone in which one set of facets shows a sapphire blue, and another set a rich emerald green. Crystals of this character have not been noticed in any other tourmaline mine in southern California, although fine blues and greens exist in other places. With the lepidolite is a granular blue and lavender mineral which could not be determined, but apparently is a lithia compound.

In examining the ledge, $2\frac{1}{2}$ feet of pegmatite were found overlying the pocket stratum. The pockets themselves were filled with soil and foreign matter, rendering it impossible to say exactly what the nature of the softer material that once filled them had been. Some pockets were hollow, containing nothing but quartz crystals, while near them were pockets absolutely filled with tourmaline pencils. The lower strata or line-rock of these ledges is also pegmatitic, although of much finer crystallization than the top. About 250 feet southwest of the tourmaline workings, the ledge is badly broken and shows only in places, in the nature of blowouts of pegmatite and quartz. In some of these blowouts golden and aquamarine beryl were found frozen in the formation. Many of these pieces were of excellent gem quality, and the owners signify their intention of doing considerable development work at these places. About $300 has been expended, producing about 20 pounds of tourmaline of gem quality. As in many other cases of prospecting and mining for gems in southern California, lack of funds has greatly hindered the proper development and exploiting of this mine.

There is neither timber nor water on the mine, but an abundance of timber can be secured within half a mile. Also, water can be piped to the property from springs on the hill above.

The mine next described does not furnish either gem-tourmaline or kunzite, but is worked as a beryl mine, some fine material having been obtained. It lies about half way between Pala and Mesa Grande, on Palomar Mountain, which is a spur or foothill of the Smith Mountain ridge, with which Pala Mountain is closely related, and hence it is considered here.

The Mack Mine.—Located at Rincon, San Diego County, in Sec. 25, T. 10 S., R. 1 W., S. B. M. This mine was discovered in November, 1903, by Mr. J. M. Mack and an Indian named J. Calec, near the Rincon Indian Reservation, at an altitude of 1960 feet (aneroid). The mine is on the Pala belt, 9½ miles southeast of Pala; the ledge has a dip of 45 degrees to the southwest, and is exposed on the hanging wall for about 75 feet. Work has been entirely confined to the surface, and but little gem material has been taken out, although several pounds of peculiar opaque, deep-blue beryl were extracted. These crystals are different from any yet found in San Diego County, and should be analyzed. Mr. Mack contemplates a great deal of development work, however, and during 1905 it will be possible to determine whether or not this locality will produce the emerald, as indications are very favorable.

The ledge is essentially pegmatite, with an average width of from

ILL. No. 38. MACK BERYL MINE, RINCON, SAN DIEGO COUNTY—VIEW FROM THE SOUTH, SHOWING MEN AT WORK TAKING OUT BERYL AT THE TOP OF THE LEDGE.

5 to 6 feet, with a gray granite foot wall. The hanging wall was hard to determine, as a great deal of matter from the ledge had fallen down and covered it at nearly every place, but was apparently a blue and gray diorite. The pockets are very narrow and are confined exclusively to a bony crystallization of orthoclase, and most of the beryls found were frozen into this crystallization. Wherever a pocket was found in which clays or other soft substances were the matrix, the crystals were exceptionally fine and could be cut into perfect gems.

So little work has been done that it is hardly of importance to report this locality if it were not for the peculiarity of the crystals found.

The owners would not give the valuation of their specimens or cut stones sold, hence the product can not be estimated. There is a small spring of water on the property, and some sycamore and oak timber. The exact locality is 1¼ miles north of the Rincon store, in the first cañon east of said store.

Since the preceding data were collected, fine gem beryls have been obtained at this mine, which are referred to in the body of this Bulletin.

South of all these localities, lies a separate group of occurrences of garnet, with beryl and in some cases topaz, centering around Ramona,

ILL. No. 39. MACK BERYL MINE, RINCON, SAN DIEGO COUNTY—PORTION OF
LEDGE, ABOUT EIGHT FEET THICK, SHOWING COARSE PEGMATITE
ABOVE AND "LINE-ROCK" BELOW.

and also the garnet country far to the southeastward in the vicinity of Jacumba. These suggest a parallel line or belt of garnet and beryl, southwest of the tourmaline-kunzite line and parallel to it; but it is not possible yet to say how far this idea may be correct. The facts, as thus far known, are as follows: The garnets belong to the variety essonite, mainly, although many of them are called spessartite (manganese garnet); but the writer is not satisfied that this latter species really occurs. Both varieties are often called hyacinth by jewelers, and may present, as at many of these points, rich orange and fulvous shades between red and yellow.

RAMONA DISTRICT.

A B C Mine.—The A B C mine, owned by Mr. Henry Daggett, of San Diego, and Mr. Alex. McIntosh, of Ramona, was discovered November 1, 1903, and is situated in the S. W. ¼ of the N. W. ¼ of Sec. 8, T. 13 S., R. 1 E., S. B. M., and at an altitude of 1950 feet. The property was acquired by location by the above-named parties on government land. It is about 4 miles northeast of Ramona, San Diego County, which is the nearest base of supplies. The vein has an average width of 7 feet,

ILL. No. 40. RAMONA DISTRICT, SAN DIEGO COUNTY—GENERAL VIEW OF LEDGES, LOOKING NORTH. TOPAZ, GARNET, BERYL, AND TOURMALINE.

and runs north 35 degrees west, with a dip of 12 degrees to the southwest. The claim embraces two ledges, very promising in character.

Three places have been opened on the ledge at the eastern extremity of the claim, and at intervals of about 50 feet. The first two are in the nature of open cuts, in which the ordinary scalping process was employed, and gems taken from broken ledge matter and soil. The principal working, however, consists of a tunnel 18 feet long, from which a stope following the pay shoot for 45 feet has been run. The work is very crude, and no system seems to have been employed in the mining.

Both foot and hanging walls are of a gray decomposed diorite, in which the feldspar has been much altered, and some quartz and biotite were found. The ledge is essentially 3 feet of coarse, poorly crystallized pegmatite, stained in some places with iron and manganese. Many black tourmaline crystals with terminals pointing directly toward the pockets were observed, somewhat altered to quartz and muscovite. Below the pegmatite is a stratum varying in width from 1 to 6 inches, composed of a grayish or whitish decomposed orthoclase, with disseminated crystals of muscovite having a pinkish and lavender tinge on the outer edges. It is in this stratum, coated with albite and clay, that the

ILL. No. 41. A B C MINE, RAMONA, SAN DIEGO COUNTY—LOOKING NORTHEAST. PINK BERYL.

pink beryls are found, generally solitary in a pocket, with two or three large blackish-green tourmaline crystals. Quartz crystals were observed both on the top and bottom of this stratum, but not in the pockets with the beryls. It was also noticed that the pink-tinged muscovite was not in contact with the beryl crystals. Contrary to what is usual in ledges of this character, the edges of the pockets do not touch between the upper and lower strata, but continue through the entire working without interruption, although widening and narrowing in places. No other minerals were found existing in the same pocket (or rounded mass of clay and decomposed spar).

Underlying this beryl-bearing stratum is about 18 inches of a soft albite, angular in crystallization, and with numerous holes penetrating the mass. In these cavities minute essonite garnets were seen, also spessartite (?) and hundreds of serrated black tourmalines, penetrating in every direction. No gems, however, were found among these. This stratum of albite lies frozen to the line-rock, or micaless granite, constituting the base of the ledge. The line-rock is coarse, and shows less interlineations than at any other mine so far observed in this locality. In places, large portions of graphic granite occur, embedded in the upper strata of ordinary pegmatite. In this graphic granite small cavities were noticed, containing steatite and montmorillonite, with lithia mica occurring at intervals. Very minute whitish crystals were found in these talcs, which appeared to be topaz, although too small for identification. In some places, also, where quartz crystals were found, disseminated crystals of pink muscovite occur, embedded and penetrating.

The minerals noted were pink beryl; green, dark green, and black tourmaline crystals; essonite and the so-called spessartite, sparingly, lepidolite, muscovite, and biotite micas, albite and orthoclase feldspars, montmorillonite, steatite, kaolin, and stains of manganese and iron.

Giant powder was used exclusively. Altogether about $500 has been expended, producing several pounds of pink beryl, the exact amount not being available at present. Some of these stones have been cut by local lapidaries, and show a rose-petal pink. They possess considerable brilliancy, and are remarkably free from hairs, flaws, or bubbles. One cut stone, weighing 30 carats, and without a flaw, was obtained from this mine.

Little Three Mine.—This mine is owned by Mr. Dan McIntosh, of Ramona, Mr. H. W. Robb, of Escondido, and Mr. Chas. F. Schnack. This prospect was discovered in May, 1903, by Mr. Robb, who had secured a permit to prospect on land owned by Messrs. McIntosh and Ferguson. It is situated in the N. E. ¼ of S. E. ¼ of Sec. 8, T. 13 S., R. 2 E., S. B. M., and is about 4½ miles northeast of Ramona, which is the nearest source of supplies.

The vein runs northeast and southwest, at an angle of north 35 degrees west, and dips to the south at an angle of 20 degrees. The average width of the vein is 4½ feet; the altitude is 1940 feet. The work so far consists of open cuts; the vein being naturally exposed for about 60 feet on the hanging wall, it has been possible to commence work where the vein enters the ground, and break open the ledge of the pegmatite to where the pockets occur in the center. About 60 square yards of the vein have been uncovered in this manner, showing some very interesting conditions of formation. At the southeast extremity of the workings, "spessartite" garnet was encountered, associated with

ILL. No. 42. LITTLE THREE MINE, RAMONA, SAN DIEGO COUNTY—SORTING TOPAZ, BERYL, AND ESSONITE GARNET.

ILL. No. 43. LITTLE THREE MINE, RAMONA, SAN DIEGO COUNTY—LOWER WORKINGS AT EAST END. TOPAZ BERYL, AND ESSONITE GARNET.

small green beryls in pockets of decomposed albite, orthoclase, and muscovite mica. In this portion of the ledge no tourmalines of any color, nor any topaz, were found in the pockets; but black tourmalines occurred very thickly interspersed in the upper or pegmatite portion of the ledge at this place. Also, the "line-rock," or micaless granite, forming the base of the ledge has parallel, wave-like bands of minute black tourmaline. The associated minerals at this part were only beryl and quartz crystals, and the beryl very sparing. A concentric band of hematite and ferruginous quartz seems to separate this particular pocket from the other pocket material found in the ledge.

From this pocket, working northwest, a gradual change was encountered and a barren condition for about 10 feet. Then coarse, bone-like concretions of albite were first discovered, with large and perfect quartz crystals. The interior of the pockets lying with these minerals has either been decomposed completely and washed away, or else the pockets were hollow, without any matter filling them, as they are at present filled with the soil, which seems to be the same as that found on the hillside above the ledge. In this loose soil, and "frozen" to the albite and orthoclase, are numerous wedge-shaped crystals of topaz, some of which weigh over a pound; they are white, sea-green, sky-blue, and light yellow in color. Attached to the roof and floor of these cavities, and with a long root extending up into the quartz and pegmatite, are gigantic tourmaline crystals, deep green, mostly opaque, some of them 5 inches in diameter, and weighing as high as 15 pounds. Some small pencil tourmalines of a deep-green color and gem quality are found loose in the pockets, also a number of small topaz crystals that have become detached from their matrix of albite. Purple and pinkish muscovite in very large crystallizations and frozen into nuggets are also observed loose in the pockets, or attached to the albite. In most cases these crystals of mica are attached to each other at right angles, leaving angular holes, in which very perfect topaz crystals have formed.

The output of this work has been approximately 30 pounds of topaz, 50 pounds of all classes of tourmalines, and a small quantity of "spessartite" garnet. Beryl pseudomorphs after topaz were also noticed, badly checked, but of pinkish and light yellow colors. Also quartz pseudomorphs taking the crystallization of the topaz, and in cubes and rhombic prisms, are found loose in the pockets of topaz.

The ledge proper is a fine-grained granitic pegmatite, with foot and hanging walls of gray decomposed diorite. The underlying line-rock in the topaz locality assumes a banded appearance, very straight in its interlineations; it is coarser than is generally seen in ledges of this kind, and is notable for the absence of either garnet or tourmaline in any quantity, the lines or bands apparently being a stain from manganese. A little biotite was also seen. This is a very strong ledge, and can be traced without a break for over 3000 feet, with an average width of 4 feet, and in some places much wider.

No work has been performed other than that described, but the ledge shows indications of garnet for its entire length.   This mine is a westerly extension of the Surprise mine, owned by J. E. Farley, James W. Booth, and Mrs. G. M. Stone.   There is quite a quantity of fine oak and sycamore

ILL. No. 44. LITTLE THREE MINE, RAMONA, SAN DIEGO COUNTY—WALL
ROCK, SHOWING LINES OF SMALL BLACK TOURMALINE.

timber in close proximity to the mines, and a spring of water sufficient for domestic uses, but which can probably be developed for all purposes needed in mining for gems.

Giant powder has been used exclusively, and no bad results have been reported.   Pocket material has been extracted, and the gems taken out by the screening process only, and quite a quantity of small crystals of good quality were found in the tailings.

Surprise Mine.—The Surprise mine, adjoining the last, is owned by J. E. Farley, J. W. Booth, and Mrs. G. M. Stone, all of Ramona, Cal., and is situated in the N. W. ¼ of S. W. ¼ of Sec. 9, T. 13 S., R. 2 E., S. B. M. It was discovered on patented land owned by Mrs. Stone, November 1, 1903, by Mrs. Booth, who noticed a few "spessartite" garnets sticking in the pegmatite. The three persons above-named became working partners to develop the property.

The vein runs nearly due east and west, but with a slight trend to the northwest and southeast. It dips to the south at an angle of about 20 degrees, and has an average width of 3½ feet. Two places have been opened on the ledge, at intervals of about 300 feet, each showing an entirely different condition in the formations. The first is about 250 feet north of Mr. Booth's residence, which is the stage station between Foster and Julian, San Diego County. At this place the pegmatite is finely crystallized, about 18 inches in width, and lies under a hanging wall of micaceous diorite. The pegmatite contains considerable graphic granite, with greenish stains, crystallized quartz, clear and white, and muscovite of a rich grass-green. Beneath this is an average thickness of a foot and a half of decomposed albite and orthoclase and infiltrated sand and earth, with some hydrated muscovite, and black tourmalines, many of which are altered to muscovite and quartz. Disseminated through this friable mass are "spessartite" garnets, in colors from deep red to light honey-yellow, affording beautiful gems, some of which have been cut, weighing from 3 to 6 carats. About five pounds of these were taken out of a cut running along the ledge about 6 feet in depth and 18 feet long, and an average width of 4 feet. Beneath this is the usual "line-rock," or micaless granite, in which no garnets were noticed, but banded lines, 2 to 3 inches apart, of minute black tourmalines were seen; this would indicate that a higher crystallization of the ledge forces the lower into the wall or outer rock, as garnets are always found to occur in the lower rock of tourmaline ledges, while the tourmaline is found as embedded crystals in the lower rock of garnet-bearing ledges, in this locality. Some quartz crystals, which appear to have been etched by either fluorides or some other chemical compound, occur broken and disseminated with the garnets.

The second working lies east of the first, and is more compact, with gray micaceous diorite as foot and hanging walls. This pegmatite is very finely crystallized, and is stained with iron and manganese, and has serrated black tourmalines. In the center of this ledge, lying between the gray base rock and the upper pegmatite, is 6 or 8 inches of orthoclase, somewhat altered. In this orthoclase occur small pockets 2 or 3 inches in diameter, filled with fine granular ferruginous quartz. In this sand, topaz is found, usually coated with a talcose clay. Those near the surface were mostly white or colorless, while at a depth of 6 feet the color had changed to sky-blue and aquamarine-blue. About four pounds of these crystals have been extracted from a cut 20 feet long and extending 8 feet in depth of the incline of the ledge.

Several very fine pink beryls were also taken out, one 6 inches long, having three perfect sides, and 1½ inches in diameter, being the largest crystal yet found. About two pounds of pink beryl has been the output so far. The above amounts of spessartite, topaz, and beryl have been extracted at an expense of $250. Giant powder is used exclusively. No work is in progress at present, but further development is contemplated.

This mine is an extension of the Little Three mine, owned by Mr. Dan McIntosh et al., adjoining it on the northwest. These parties own several other ledges in the same vicinity, traversing four quarter-sections of land owned by them, and lying in a line extending east from the present workings.

Timber and water are available in sufficient quantities for mining purposes. The stones are extracted in both localities by screening and washing, and the owners seem to be thorough in their work.

The occurrence of a yellowish, reniform, compact and extremely heavy substance was noted in some of the topaz pockets. The specific gravity of this mineral and its peculiar color have attracted the attention of several people, but it was impossible, with the means at hand, to determine what it was. From the edge of the pockets containing these nuggets, radiated black tourmalines were found, altered to a micaceous substance of sea and emerald green color, with occasional tinges of purple and rose-pink. This alteration seems to be an allied mineral to the above. Some triplite and magnetic iron occur at the junction of the foot wall and pegmatite.

The list of minerals noticed in these mines was:—white and blue topaz; pink, green, and white beryl; black, green and brown tourmaline; spessartite (so called), biotite, magnetite, orthoclase, albite, quartz in fine crystallizations, and the two unknown minerals above referred to.

Hercules Mine.—This mine, owned by Messrs. Samuel G. Ingle and Harry Titus, of San Diego, and Mr. Pray, of Escondido, lies about 4½ miles northeast of Ramona, and about three fourths of a mile northwest of the stage station between Foster and Julian. The mine was located in August, 1903, by the above-named parties.

The work has been confined entirely to open cuts or scalping, and the gems have been extracted in all cases either from the débris or from broken pockets in the ledge, which is a coarse pegmatite, decomposed, and with very little perfection, in the albite or orthoclase; but where black tourmalines penetrate this crystallization, joining on to the black tourmaline, embedded in either albite or orthoclase, are essonite garnet and so-called spessartite.   The latter is of the finest quality, and has produced gems from one to six and eight carats in weight, without flaw, which are retailing at $20 a carat.

Both hanging and foot walls are a gray diorite, in which some mica can be found.   The course of the vein is north 60 degrees west, with a dip of 45 degrees.   The location is in the S. E. ¼ of S. E. ¼ of Sec. 6, T. 13 S., R. 2 E., S. B. M.

The method of handling the product has been confined exclusively to screening, and a good many gems have been thrown over on account of the peculiar condition of the clays which cover them; but the output

ILL. No. 45. HERCULES MINE, RAMONA, SAN DIEGO COUNTY—ESSONITE GARNET AND BERYL.

so far has been 15 pounds of garnet and a half pound of very clear green beryl, which is an associated mineral with it.   A few green and blue tourmalines, but not fit for gem purposes, have been found higher up on the ledge.

There is a spring on this property, which will furnish enough water for domestic and mining purposes; also sycamore and oak timber in sufficient quantity for mining.

Lookout Mine.—This mine, owned by Messrs. Samuel G. Ingle and Harry Titus, of San Diego, and Mr. Pray, of Escondido, was located in the month of July, 1903.   It is situated 4½ miles northeast of Ramona, and joins the Hercules mine on the northeast. The vein has a dip of 20 degrees to the southwest, and an average width of 4½ feet. The claim runs north 55 degrees west, and is located in the S. W. ¼ of Sec. 5, T. 13 S., R. 2 E., S. B. M.

Work on this claim has been confined entirely to open cuts and scalping. Garnet, called spessartite, is the only gem found, although indications of beryl and tourmaline, with two or three peculiar metallic substances which could not be determined, were noticed. Both walls are of gray diorite, containing some biotite mica, although a seam of red clay lies between either wall and the ledge itself. The ledge is composed partly of feldspar, with very little quartz. In the pockets, albite and orthoclase are the mother of crystallization, and a very peculiar condition of the quartz is evident. The crystals seem to have been broken at some time into splinters, and then welded together, forming a conglomerated mass of quartz with no distinct crystallization. Adhering to this quartz, and also to the surface of the albite, are perfectly formed garnet crystals, which, in many cases, have afforded beautiful gems.

Not enough work has been done to make a satisfactory examination. About three pounds of garnet, and perhaps four ounces of fine beryl, was the total product of this mine.

Some sycamore and oak timber are available; and water owned by the same parties on the Hercules mine can be used in connection with this one.

McFall Mine.—This mine is situated 7½ miles southwest of Ramona, on the eastern line of the San Vicente grant. It is owned by John McFall, who located the property about ten years ago as a zinc mine, and erroneous reports were given of its value as a zinc property. On examination, no zinc was found nor any indications of it, but a large body of essonite garnet and finely crystallized epidote was shown.

A shaft 22 feet in depth still remains in solid garnet, with very little impurity of quartz. Very few gems were found, however, although many handsome crystals, more or less transparent, were among those taken out. There is a certain condition in these crystals which does not produce good refraction of light, and hence as gems they have no value. The epidote, however, is the finest yet seen in San Diego County, and will probably produce gems.

Mr. McFall expects to work the property for abrasive purposes, as transportation can be secured cheap enough to make this course profitable. Both wood and water are adjacent to the property, but not on it. No work has been done on the mine for some little time.

The formations are both blue and gray diorite, and the masses of garnet appear to be pockets rather than ledges.

Prospect Mine.—This mine, owned by Messrs. H. A. Warnock and John P. Sutherland, of Ramona, was located on September 15, 1904, by the above-named parties, and is about 4 miles northwest of Ramona, crossing the road between Ramona and Mesa Grande, an open cut having been made on the east side of the road in Hatfield Cañon.

Spessartite (so called) has been the only product in gems, although greenish tourmalines have also been found. The ledge is about 6 feet in width, of a poorly crystallized pegmatite, and most of the gems are found frozen into the ledge, few pockets having been discovered.

The parties are actively at work, and probably will find a better condition in 15 or 20 feet from the present working. The output has been very small, and no sales have been made. The prospect is worth mentioning, however, as it is the last mine on the northwest end of the Ramona belt of crystallization, the belt apparently being barren for 14 miles northward of Mesa Grande.

Messrs. Warnock and Sutherland expect to continue their work until something definite is known about the property, and a report two or three months later will be more satisfactory than can be had at present.

There are both wood and water in plenty on the property. It is now owned by Mr. Warnock.

JACUMBA DISTRICT.

Dos Cabezas Mine.—This mine is 17 miles north and east from Jacumba Hot Springs by road, although in a direct line only about 8 miles; it is situated in Sec. 2, T. 17 S., R. 8 E. Here many fine hyacinth garnets have been taken out from a matrix of carbonate of lime, which occurs in quantities sufficient to be used as building marble, etc. There are also indications of phosphate of lime. This locality has been worked off and on for the last ten years for gem crystals, and several hundred dollars' worth have been extracted by different parties, but nothing definite has been done, owing to its inaccessibility and the lack of wood and water. Properties now owned in that vicinity are those of Mr. James Jasper, the San Diego Desert Marble Company, the San Diego Gem Company, W. H. Trenchard, T. H. Steinmeyer, and William Hill. Development is expected during the next year.

Nine and a half miles east of Jacumba, and near Mountain Springs, on the road leading from San Diego to Imperial, and on unsurveyed land, is a locality on which three or four prospects have been located showing excellent essonite and so-called spessartite garnet. These localities are now controlled by the San Diego Gem Company, 1529 D street, San Diego, who have sunk a shaft and have done considerable surface work. The gems extracted are of exceptional quality and size. Several thousand dollars will be expended by the owners during the

ILL. No. 46. PROSPECT MINE, RAMONA, SAN DIEGO COUNTY—GENERAL VIEW.
ESSONITE GARNET.

ILL. No. 47. PROSPECT MINE, RAMONA, SAN DIEGO COUNTY—SHOWING LEDGE AND
OPENING. ESSONITE GARNET.

next year. The water supply is about 4½ miles away, and there is no timber whatever or even wood for ordinary purposes. The country is very rough and inaccessible, but bids fair to be one of the best producers of gems yet discovered in California.

Crystal Gem Mine.—This is owned by Collier & Smith, of San Diego, and is situated about 8½ miles northwest from Jacumba. Pink and green beryls associated with essonite and (so-called) spessartite garnet have been the only output, but general indications are very favorable. The ledge is a coarse pegmatite, about 8 feet in width, and extends for nearly a mile. Quartz crystals, albite, orthoclase, and indications of lithia are also found in conjunction. This property is not worked at present, but probably will be during the next year. Ten pounds of fine essonite garnet and perhaps three or four pounds of beryl were taken out during 1904. There is a spring of water on the property, and plenty of timber.

Manganese Deposits.—These are owned by the San Diego Desert Marble Company, and lie 1½ miles northwest of Jacumba Hot Springs. A ledge averaging 10 feet in width and extending about 5000 feet has been located by these parties, and shows oxides of manganese associated with garnet, beryl, and black tourmaline. No development work has been done, but upon the advent of a railroad this property may become valuable, as the manganese is of exceptional quality and can be utilized in many ways.

Farther to the north and east are other localities, in the vicinity of Seventeen Palms, in the Santa Rosa Mountains, on the edge of the desert, and in the direction of Salton Lake, where Mr. H. C. Gordon reports fine and abundant occurrences of garnet. Much of this is the wild and barren region claimed by the old Indian chief known as "Fig-tree John," elsewhere mentioned.

TURQUOISE—SAN BERNARDINO COUNTY.

Turquoise mining is carried on by two principal companies in San Bernardino County—the Toltec and the Himalaya.

Toltec Gem Mining Company.—The California property of this company consists of three groups of mines situated in San Bernardino County on the Great Desert about 100 miles northwest of Needles, and about 50 miles north from Manvel, which is on a branch of the Santa Fé Railroad. The altitude is between 5000 and 6000 feet; and there being no water at either of the camps, it is necessary to draw it over the mountains from 1 to 5 miles. These camps are about 6 miles apart, and are known as East Camp, Middle Camp, and West Camp, in the old Solo Mining District. Death Valley is within 20 miles of West Camp. These mines are all patented. The qualities of the turquoise taken from these various camps vary widely, from quite soft to very hard. The same company also has turquoise mines in Nevada, 60 miles due east of these. Here stone hammers were met with at a depth of 18 feet. Scarcely any turquoise was found much below 100 feet from the surface, and a 200-foot shaft failed to reveal any at all. This fact, which is also reported from the mines of the Himalaya Company, is a curious one, indicating that the turquoise must be in some way a product of rather superficial alteration. The mines of both these companies have been quite large producers. The Toltec Company obtained one gem-stone, of rather a pale blue, that cut into a perfect oval measuring 32 by 45 millimeters, and weighing 203 carats.

Himalaya Mining Company.—This company also has been operating turquoise mines in San Bernardino County. They are owned by Mr. L. Tannenbaum, of New York, and are likewise situated in the Solo Mining District, but 60 miles due west of Manvel, where a team must be secured to visit the mine, as no other means of communication exists. There are five claims in this group, all of them on the same ledge, which consists of bird's-eye porphyry with some granite, with a north and south strike and a dip of 75 degrees west. The pockets of turquoise, which is practically the only gem found, lie in this porphyry surrounded by a friable mass of so-called silicate of lime. Two shafts 80 feet in depth have been sunk on the property, but 40 feet was the lowest level at which gem-turquoise was found. From this level the mine has been practically stoped to the surface. There is no timber at the mine, but some small pine can be secured 10 miles north. Water was found by sinking a well 85 feet; it can be used for drinking purposes, but is of very inferior quality. Work was done by the screening and washing method, and entirely by hand, there being no machinery of any kind on the property. Other improvements are bunk-houses, etc. The mine was closed down on the first day of March, 1903, and no work of any account has been done since that date. For the two months during which the mine was worked in that year, six men were employed, at an average wage of $2.50 per day. The expense of mining was very high—about $20 per foot. Giant powder was used exclusively, and it required about $10\frac{1}{2}$ pounds to the foot.

The shipments that year (January and February, 1903), as given by the Wells-Fargo agent, amounted to 431 pounds of matrix and ordinary turquoise, and 49 pounds of picked material.

CHRYSOPRASE—TULARE COUNTY.

Chrysoprase was discovered in Tulare County in 1878, by Mr. George W. Smith, a surveyor of Visalia. He presented specimens to Mr. M. Braverman of Visalia, who identified them as chrysoprase. Later, the State Mining Bureau confirmed this determination. The first specimen was the finest ever found in the county; it was sent to the Paris Exposition of 1889, but failed to reach its destination, being stolen en route. It was about 3½ inches long and 1½ inches thick. The first discovery, and thus far the best outcroppings which have been developed, are located on the McGinnis property, in the N. E. ¼ of Sec. 8, T. 18 S., R. 26 E., Mount Diablo meridian. This location is about 10 miles northeast of Visalia, on Venice Hill. These outcroppings extend along the southeastern slope of this hill and through Section 8 and the S. W. ¼ of Section 4. Here it occurs in small irregular veins, which range from mere seams to veins 2 and 3 inches in thickness, in a somewhat altered red jasper rock.

When visited by the Field Assistant of the Mining Bureau (February, 1905) the larger veins had been deprived of all the chrysoprase suitable for specimens, but the excavations show that the chalcedony veins persist, indicating the structure of the former deposits, and also the gradation from the darker green chrysoprase of the outcrops through a lighter green (prase-opal) until at a depth of about 20 feet a pure-white milky quartz or chalcedony is encountered, which is free from all nickel oxide coloring.

As to the size and quality of the gems taken out, we find in the seventeenth annual report of the U. S. Geological Survey[*] the following note: "It is much flawed and good pieces for cutting are scarce, but the color is excellent, and some handsome articles of small size have been made from it." Since then many fine stones have been found.

The chrysoprase outcroppings have been traced at different places all along the foothills of the Sierra Nevada across Tulare County, and the following locations have been noted by different authorities: Venice Hill, Stokes Mountain, Tule River, Deer Creek, and one mile east of Lindsay. All of these have been announced in the annual reports of the writer on the production of precious stones in the United States, for 1895 to 1898 inclusive. Of these, the Himalaya Mining Company owns three claims at the chief locality at Venice Hill, a short distance northeast of Visalia; also two at Lindsay, a little northeast and southeast of the town, respectively; and one on Deer Creek, in Sec. 20, T. 22 S., R. 28 E. One or two other large claims at Venice Hill, another some distance east of Visalia, in the N. W. ¼ of Sec. 28, T. 18 S., R. 27 E., and

---

[*] Min. Res. U. S., Rept. U. S. Geol. Survey, 1895, p. 912.

one or two in the Stokes Mountain region, Secs. 9 and 10, T. 16 S., R. 26 E., Mount Diablo base and meridian, belong to other parties.

For a time a very active interest was taken in mining chrysoprase, but of late years there has been less demand than formerly, because this stone is not now the fad in jewelry that it was at first.

No ruling price or valuation can be set on the rough stone, as the character of a specimen can not be told with any degree of certainty until the piece is cut. The cut stones brought prices from $2 to $3 per carat. Most of the Tulare County output was cut in New York, but some few stones were cut in San Francisco and Denver.

CALIFORNITE (MASSIVE VESUVIANITE)—SISKIYOU COUNTY.

The exact location of the deposit of californite, as the writer has proposed to call the compact, jade-like variety of vesuvianite, or idocrase, described in full on page 93 of this Bulletin, is given as follows by Mr. Edwin L. Hoyt of Yreka, in a recent communication:

"The californite deposit in Siskiyou County is 10 miles north of Happy Camp, which is the nearest postoffice to the property; also 10 miles north of Klamath River. The Happy Camp placer mining district parallels the river in this section. The nearest railroad station is that at Yreka, 90 miles distant by wagon road. I am unable to give dip of vein, etc., but believe it is more of a deposit or kidney or large lens, as there are no indications of a fissure."

APPENDIX.

ILL. No. 48. FERRY BUILDING, SAN FRANCISCO, ONE HALF THE UPPER FLOOR OF WHICH IS OCCUPIED BY THE STATE MINING BUREAU.

CALIFORNIA STATE MINING BUREAU.

This institution aims to be the chief source of reliable information about the mineral resources and mining industries of California.

It is encouraged in its work by the fact that its publications have been in such demand that large editions are soon exhausted. In fact, copies of them now command high prices in the market.

The publications, as soon as issued, find their way to the scientific, public, and private libraries of all countries.

STATE MINERALOGIST.

The California State Mining Bureau is under the supervision of Hon. Lewis E. Aubury, State Mineralogist.

It is supported by legislative appropriations, and in some degree performs work similar to that of the geological surveys of other States; but its purposes and functions are mainly practical, the scientific work being clearly subordinate to the economic phases of the mineral field, as shown by the organic law governing the Bureau, which is as follows:

SEC. 4. It shall be the duty of said State Mineralogist to make, facilitate, and encourage special studies of the mineral resources and mineral industries of the State. It shall be his duty: To collect statistics concerning the occurrence of the economically important minerals and the methods pursued in making their valuable constituents available for commercial use; to make a collection of typical geological and mineralogical specimens, especially those of economic or commercial importance, such collection constituting the Museum of the State Mining Bureau; to provide a library of books, reports, drawings, bearing upon the mineral industries, the sciences of mineralogy and geology and the arts of mining and metallurgy, such library constituting the Library of the State Mining Bureau; to make a collection of models, drawings, and descriptions of the mechanical appliances used in mining and metallurgical processes; to preserve and so maintain such collections and library as to make them available for reference and examination, and open to public inspection at reasonable hours; to maintain, in effect, a bureau of information concerning the mineral industries of this State, to consist of such collections and library, and to arrange, classify, catalogue, and index the data therein contained, in a manner to make the information available to those desiring it, and to provide a custodian specially qualified to promote this purpose; to make a biennial report to the Board of Trustees of the Mining Bureau, setting forth the important results of his work, and to issue from time to time such bulletins as he may deem advisable concerning the statistics and technology of the mineral industries of this State.

THE BULLETINS.

The field covered by the books issued under this title is shown in the list of publications. Each bulletin deals with only one phase of mining. Many of them are elaborately illustrated with engravings and maps. Only a nominal price is asked, in order that those who need them most may obtain a copy.

THE REGISTERS OF MINES.

The Registers of Mines form practically both a State and a County directory of the mines of California, each county being represented in a separate pamphlet. Those who wish to learn the essential facts about any particular mine are referred to them. The facts and figures are given in tabular form, and are accompanied by a topographical map of the county on a large scale, showing location of each mineral deposit, towns, railroads, roads, power lines, ditches, etc.

HOME OF THE BUREAU.

The Mining Bureau occupies the north half of the third floor of the Ferry Building, in San Francisco. All visitors and residents are invited to inspect the Museum, Library, and other rooms of the Bureau and gain a personal knowledge of its operations.

THE MUSEUM.

The Museum now contains over 16,000 specimens, carefully labeled and attractively arranged in showcases in a great, well-lighted hall, where they can be easily studied. The collection of ores from California mines is of course very extensive, and is supplemented by many cases of characteristic ores from the principal mining districts of the world. The educational value of the exhibit is constantly increased by substituting the best specimens obtainable for those of less value.

These mineral collections are not only interesting, beautiful, and in every way attractive to the sightseers of all classes, but are also educational. They show to manufacturers, miners, capitalists, and others the character and quality of the economic minerals of the State, and where they are found. Plans have been formulated to extend the usefulness of the exhibit by special collections, such as one showing the chemical composition of minerals; another showing the mineralogical composition of the sedimentary, metamorphic, and igneous rocks of the State; the petroleum-bearing formations, ore bodies, and their country rocks, etc.

Besides the mineral specimens, there are many models, maps, photographs, and diagrams illustrating the modern practice of mining, milling, and concentrating, and the technology of the mineral industries. An educational series of specimens for high schools has been inaugurated, and new plans are being formulated that will make the Museum even more useful in the future than in the past. Its popularity is shown by the fact that over 100,000 visitors registered last year, while many failed to leave any record of their visit.

ILL. No. 49. MINERAL MUSEUM, CALIFORNIA STATE MINING BUREAU.

THE LIBRARY.

This is the mining reference library of the State, constantly consulted by mining men, and contains between 4000 and 5000 volumes of selected works, in addition to the numerous publications of the Bureau itself. On its shelves will be found reports on geology, mineralogy, mining, etc., published by states, governments, and individuals; the reports of scientific societies at home and abroad; encyclopædias, scientific papers, and magazines; mining publications; and the current literature of mining ever needed in a reference library.

Manufacturers' catalogues of mining and milling machinery by California firms are kept on file. The Registers of Mines form an up-to-date directory for investor and manufacturer.

The librarian's desk is the general bureau of information, where visitors from all parts of the world are ever seeking information about all parts of California.

READING-ROOM.

This is a part of the Library Department and is supplied with over one hundred current publications. Visitors will find here various California papers and leading mining journals from all over the world.

The Library and Reading-Room are open to the public from 9 A. M. to 5 P. M. daily, except Sundays and holidays.

THE LABORATORY.

This department identifies for the prospector the minerals he finds, and tells him the nature of the wall rocks or dikes he may encounter in his workings; but this department *does not* do assaying nor compete with private assayers. The presence of minerals is determined, but not the percentage present. No charges for this service are made to any resident of the State. Many of the inquiries made of this department have brought capital to the development of new districts. Many technical questions have been asked and answered as to the best chemical and mechanical processes of handling ores and raw material. The laboratory is well equipped.

THE DRAUGHTING-ROOM.

In this room are prepared scores of maps, from the small ones filling only a part of a page, to the largest County and State maps; and the numerous illustrations, other than photographs, that are constantly being required for the Bulletins and Registers of Mines. In this room, also, will be found a very complete collection of maps of all kinds relating to the industries of the State, and one of the important duties of the department is to make such additions and corrections as will keep the maps up to date. The seeker after information inquires here if he wishes to know about the geology or topography of any district; about the locations of the new camps, or positions of old or abandoned

ILL. No. 50. LIBRARY AND FREE READING ROOM, CALIFORNIA STATE MINING BUREAU.

ones; about railroads, stage roads, and trails; or about the working drawings of anything connected with mining.

MINERAL STATISTICS.

One of the features of this institution is its mineral statistics. Their annual compilation by the State Mining Bureau began in 1893. No other State in the Union attempts so elaborate a record, expends so much labor and money on its compilation, or secures so accurate a one.

The State Mining Bureau keeps a careful, up-to-date, and reliable but confidential register of every producing mine, mine-owner, and mineral industry in the State. From them are secured, under pledge of secrecy, reports of output, etc., and all other available sources of information are used in checking, verifying, and supplementing the information so gained. This information is published in an annual tabulated, statistical, single-sheet bulletin, showing the mineral production by both substances and counties.

TOTAL GOLD PRODUCT OF CALIFORNIA—1848–1903.

Year	Amount	Year	Amount	Year	Amount	Year	Amount
1848	$245,301	1863	$23,501,736	1878	$18,839,141	1893	$12,422,811
1849	10,151,360	1864	24,071,423	1879	19,626,654	1894	13,923,281
1850	41,273,106	1865	17,930,858	1880	20,030,761	1895	15,334,317
1851	75,938,232	1866	17,123,867	1881	19,223,155	1896	17,181,562
1852	81,294,700	1867	18,265,452	1882	17,146,416	1897	15,871,401
1853	67,613,487	1868	17,555,867	1883	24,316,873	1898	15,906,478
1854	69,433,931	1869	18,229,044	1884	13,600,000	1899	15,336,031
1855	55,485,395	1870	17,458,133	1885	12,661,044	1900	15,863,355
1856	57,509,411	1871	17,477,885	1886	14,716,506	1901	16,989,044
1857	43,628,172	1872	15,482,194	1887	13,588,614	1902	16,910,320
1858	46,591,140	1873	15,019,210	1888	12,750,000	1903	16,471,264
1859	45,846,599	1874	17,264,836	1889	11,212,913		
1860	44,095,163	1875	16,876,009	1890	12,309,793	Total	$1,395,746,672
1861	41,884,995	1876	15,610,723	1891	12,728,869		
1862	38,854,668	1877	16,501,268	1892	12,571,900		

COUNTY RANK IN GOLD PRODUCT IN 1903.

While gold is still the leading mining product, its yield no longer puts the greatest gold-producing county in the first place. The petroleum of Kern County and the copper of Shasta give them precedence. Gold is more widely distributed than any other substance thus far mined in California; 34 counties out of the 57 in the State showing a gold yield in 1903, and it is known to exist in several others. The order in rank of the counties of the State, in the production of gold alone, is at present as follows:

1. Nevada	$2,458,047	13. Plumas	$424,112	25. Fresno	$21,538
2. Calaveras	1,904,125	14. S. Bernardino	381,197	26. Riverside	13,453
3. Tuolumne	1,732,572	15. Sacramento	335,646	27. Tulare	9,215
4. Amador	1,609,744	16. Mono	334,713	28. Monterey	8,920
5. Butte	1,571,507	17. Sierra	310,770	29. Los Angeles	8,674
6. Kern	1,022,353	18. El Dorado	277,304	30. Del Norte	7,183
7. Shasta	771,242	19. Yuba	125,830	31. Alpine	2,701
8. Siskiyou	613,576	20. Madera	93,070	32. San Luis Obispo	1,840
9. Trinity	607,728	21. Lassen	91,102	33. Ventura	1,087
10. Placer	570,571	22. Inyo	66,045	34. Orange	150
11. Mariposa	542,355	23. Stanislaus	52,869		
12. San Diego	461,516	24. Humboldt	38,509	Total	$16,471,264

ILL. No. 51.   LABORATORY, CALIFORNIA STATE MINING BUREAU.

(165)

TOTAL MINERAL PRODUCT OF CALIFORNIA FOR 1903.

The following table shows the yield and value of mineral substances of California for 1903, as per returns received at the State Mining Bureau, San Francisco, in answer to inquiries sent to producers:

	Quantity.	Value.
Asphalt	41,670 tons	$503,659
Bituminous Rock	21,944 "	53,106
Borax (Crude)	34,430 "	661,400
Cement	640,868 bbls.	968,727
Chrome	150 tons	2,250
Chrysoprase	....	500
Clays: For Pottery	90,972 tons	99,907
For Brick	214,403 M	1,999,546
Coal	93,026 tons	265,383
Copper	19,113,861 lbs.	2,520,997
Fuller's Earth	250 tons	4,750
Glass Sand	7,725 "	7,525
Gold	....	16,471,264
Granite	408,625 cu. ft.	678,670
Gypsum	6,914 tons	46,441
Infusorial Earth	2,703 "	16,015
Lead	110,000 lbs.	3,960
Lime	496,587 bbls.	418,280
Limestone	125,919 tons	163,988
Lithia Mica	700 "	27,300
Macadam	605,185 "	436,172
Manganese	1 "	25
Magnesite	1,361 "	20,515
Marble	84,624 cu. ft.	97,354
Mica	50 tons	3,800
Mineral Paint	2,370 "	3,720
Mineral Water	1,978,340 gals.	558,201
Natural Gas	120,134 M cu. ft.	75,237
Paving Blocks	4,854 M	134,642
Petroleum	24,340,839 bbls.	7,313,271
Platinum	....	1,052
Pyrites	24,311 tons	94,000
Quartz Crystals	....	1,968
Quicksilver	32,094 flasks	1,335,954
Rubble	1,610,440 tons	1,237,419
Salt	102,895 "	211,365
Sandstone	353,002 cu. ft.	585,309
Serpentine	99 " "	800
Silver	....	517,444
Slate	10,000 squares	70,000
Soapstone	219 tons	10,124
Soda	18,000 "	27,000
Tourmaline	....	100,000
Turquoise	....	10,000
Total value		$37,759,040

RELATIVE RANK OF COUNTIES IN TOTAL MINERAL PRODUCT IN 1903.

1. Kern	$4,957,602	20. Mariposa	$552,516	39. Inyo	$139,563
2. Shasta	3,201,680	21. Alameda	530,207	40. Yuba	125,871
3. Los Angeles	2,549,128	22. Sacramento	506,796	41. Lassen	92,305
4. Nevada	2,466,044	23. Madera	489,525	42. Stanislaus	70,605
5. Calaveras	2,270,668	24. Riverside	446,449	43. Contra Costa	62,500
6. Tuolumne	1,791,056	25. Plumas	424,894	44. Monterey	51,436
7. Amador	1,639,819	26. Colusa	420,468	45. Humboldt	49,316
8. Butte	1,581,325	27. Solano	404,614	46. San Joaquin	44,489
9. San Bernardino	1,516,618	28. Santa Barbara	384,688	47. Tulare	41,175
10. Orange	1,029,435	29. San Benito	367,851	48. Kings	24,200
11. Napa	896,848	30. Mono	360,024	49. Mendocino	20,580
12. Fresno	848,628	31. Sierra	311,246	50. Del Norte	7,183
13. San Francisco	802,786	32. Lake	294,018	51. Tehama	7,000
14. Placer	800,985	33. El Dorado	284,304	52. Alpine	2,847
15. Ventura	714,766	34. S'n Luis Obispo	257,416	53. Yolo	144
16. Santa Clara	670,159	35. Santa Cruz	254,247	54. Merced	780
17. Siskiyou	663,598	36. San Mateo	252,500	Unapportioned	377,783
18. San Diego	663,315	37. Marin	218,427		
19. Trinity	621,244	38. Sonoma	195,369	Total	$37,759,040

MINING BUREAU PUBLICATIONS.

Publications of this Bureau will be sent on receipt of the requisite amount and postage. Only stamps, coin or money orders will be accepted in payment. (*All publications not mentioned are exhausted.*)

Attention is respectfully called to that portion of Section 8, amendment to the Mining Bureau Act, approved March 10, 1903, which states:

"The Board (Board of Trustees) is hereby empowered to fix a price upon, and to dispose of to the public, at such price, any and all publications of the Bureau, including reports, bulletins, maps, registers, etc. The sum derived from such disposition must be accounted for and used as a revolving printing and publishing fund for other reports, bulletins, maps, registers, etc. The prices fixed must approximate the actual cost of printing and issuing the respective reports, bulletins, maps, registers, etc., without reference to the cost of obtaining and preparing the information embraced therein."

	Price.	Postage.
Report XI—1892, First Biennial	$1 00	$0 15
Report XIII—1896, Third Biennial	1 00	20
Bulletin No. 5—"Cyanide Process" (4th edition), bound	60	08
Bulletin No. 6—"Gold Mill Practices in California" (3d edition)	50	04
Bulletin No. 9—"Mine Drainage, Pumps, etc.," bound	60	08
Bulletin No. 15—"Map of Oil City Oil Fields, Fresno County, Cal."	05	02
Bulletin No. 16—"Genesis of Petroleum and Asphaltum in California," (3d edition)	30	03
Bulletin No. 18—"Mother Lode Region in California"	35	06
Bulletin No. 23—"Copper Resources of California"	50	12
Bulletin No. 24—"Saline Deposits of California"	50	10
Bulletin No. 27—"Quicksilver Resources of California"	75	08
Bulletin No. 30—"Bibliography Relating to the Geology, Palæontology and Mineral Resources of California, including List of Maps"	50	10
Bulletin No. 31—"Chemical Analyses of California Petroleum"	---	02
Bulletin No. 32—"Production and Use of California Petroleum"	75	08
Bulletin No. 33—"Mineral Production of California—1903"	---	02

	Price	Postage
Bulletin No. 34—"Mineral Production of California for 17 Years"	...	$0 02
Bulletin No. 35—"Mines and Minerals of California"	...	04
Bulletin No. 36—"Gold Dredging in California"	$0 50	06
Map of Mother Lode	05	02
Gold Production in California from 1848 to 1904	...	02
Register of Mines, with Map, Plumas County	25	08
Register of Mines, with Map, Siskiyou County	25	08
Register of Mines, with Map, Trinity County	25	08
Register of Mines, with Map, Lake County	25	08
Register of Mines, with Map, Nevada County	25	08
Register of Mines, with Map, Placer County	25	08
Register of Mines, with Map, El Dorado County	25	08
Register of Mines, with Map, Inyo County	25	08
Register of Mines, with Map, Shasta County	25	08
Register of Mines, with Map, San Bernardino County	25	08
Register of Mines, with Map, San Diego County	25	08
Register of Mines, with Map, Sierra County	25	08
Register of Mines, with Map, Amador County	25	08
Register of Mines, with Map, Tuolumne County	25	08
Register of Mines, with Map, Butte County	25	08
Register of Mines, with Map, Mariposa County	25	08
Register of Mines, with Map, Kern County	25	08
Register of Oil Wells, with Map, Los Angeles City	35	02
Relief and Mineral Map of California	25	05
Map of Calaveras County	25	08

In Preparation:

Structural and Industrial Materials of California.

Samples of any mineral found in the State may be sent to the Bureau for identification, and the same will be classified free of charge. It must be understood, however, that *no assays, or quantitative determinations, will be made.* Samples should be in lump form if possible, and the outside of package should be marked plainly with name of sender, postoffice address, etc., and a *stamp* should be inclosed for reply.

INDEX.

GOLD RUSH BOOKS

OREGON, USA

www.GoldMiningBooks.com

Books On Mining

Visit: www.goldminingbooks.com to order your copies or ask your favorite book seller to offer them.

Mining Books by Kerby Jackson

<u>Gold Dust: Stories From Oregon's Mining Years</u> - Oregon mining historian and prospector, Kerby Jackson, brings you a treasure trove of seventeen stories on Southern Oregon's rich history of gold prospecting, the prospectors and their discoveries, and the breathtaking areas they settled in and made homes. 5" X 8", 98 ppgs. Retail Price: $11.99

<u>The Golden Trail: More Stories From Oregon's Mining Years</u> - In his follow-up to "Gold Dust: Stories of Oregon's Mining Years", this time around, Jackson brings us twelve tales from Oregon's Gold Rush, including the story about the first gold strike on Canyon Creek in Grant County, about the old timers who found gold by the pail full at the Victor Mine near Galice, how Iradel Bray discovered a rich ledge of gold on the Coquille River during the height of the Rogue River War, a tale of two elderly miners on the hunt for a lost mine in the Cascade Mountains, details about the discovery of the famous Armstrong Nugget and others. 5" X 8", 70 ppgs. Retail Price: $10.99

Oregon Mining Books

<u>Geology and Mineral Resources of Josephine County, Oregon</u> - Unavailable since the 1970's, this important publication was originally compiled by the Oregon Department of Geology and Mineral Industries and includes important details on the economic geology and mineral resources of this important mining area in South Western Oregon. Included are notes on the history, geology and development of important mines, as well as insights into the mining of gold, copper, nickel, limestone, chromium and other minerals found in large quantities in Josephine County, Oregon. 8.5" X 11", 54 ppgs. Retail Price: $9.99

<u>Mines and Prospects of the Mount Reuben Mining District</u> - Unavailable since 1947, this important publication was originally compiled by geologist Elton Youngberg of the Oregon Department of Geology and Mineral Industries and includes detailed descriptions, histories and the geology of the Mount Reuben Mining District in Josephine County, Oregon. Included are notes on the history, geology, development and assay statistics, as well as underground maps of all the major mines and prospects in the vicinity of this much neglected mining district. 8.5" X 11", 48 ppgs. Retail Price: $9.99

<u>The Granite Mining District</u> - Notes on the history, geology and development of important mines in the well known Granite Mining District which is located in Grant County, Oregon. Some of the mines discussed include the Ajax, Blue Ribbon, Buffalo, Continental, Cougar-Independence, Magnolia, New York, Standard and the Tillicum. Also included are many rare maps pertaining to the mines in the area. 8.5" X 11", 48 ppgs. Retail Price: $9.99

<u>Ore Deposits of the Takilma and Waldo Mining Districts of Josephine County, Oregon</u> - The Waldo and Takilma mining districts are most notable for the fact that the earliest large scale mining of placer gold and copper in Oregon took place in these two areas. Included are details about some of the earliest large gold mines in the state such as the Llano de Oro, High Gravel, Cameron, Platerica, Deep Gravel and others, as well as copper mines such as the famous Queen of Bronze mine, the Waldo, Lily and Cowboy mines. This volume also includes six maps and 20 original illustrations. 8.5" X 11", 74 ppgs. Retail Price: $9.99

<u>Metal Mines of Douglas, Coos and Curry Counties, Oregon</u> - Oregon mining historian Kerby Jackson introduces us to a classic work on Oregon's mining history in this important re-issue of Bulletin 14C Volume 1, otherwise known as the Douglas, Coos & Curry Counties, Oregon Metal Mines Handbook. Unavailable since 1940, this important publication was originally compiled by the Oregon Department of Geology and Mineral Industries includes detailed descriptions, histories and the geology of over 250 metallic mineral mines and prospects in this rugged area of South West Oregon. 8.5" X 11", 158 ppgs. Retail Price: $19.99

Metal Mines of Jackson County, Oregon - Unavailable since 1943, this important publication was originally compiled by the Oregon Department of Geology and Mineral Industries includes detailed descriptions, histories and the geology of over 450 metallic mineral mines and prospects in Jackson County, Oregon. Included are such famous gold mining areas as Gold Hill, Jacksonville, Sterling and the Upper Applegate. **8.5" X 11", 220 ppgs. Retail Price: $24.99**

Metal Mines of Josephine County, Oregon - Oregon mining historian Kerby Jackson introduces us to a classic work on Oregon's mining history in this important re-issue of Bulletin 14C, otherwise known as the Josephine County, Oregon Metal Mines Handbook. Unavailable since 1952, this important publication was originally compiled by the Oregon Department of Geology and Mineral Industries includes detailed descriptions, histories and the geology of over 500 metallic mineral mines and prospects in Josephine County, Oregon. **8.5" X 11", 250 ppgs. Retail Price: $24.99**

Metal Mines of North East Oregon - Oregon mining historian Kerby Jackson introduces us to a classic work on Oregon's mining history in this important re-issue of Bulletin 14A and 14B, otherwise known as the North East Oregon Metal Mines Handbook. Unavailable since 1941, this important publication was originally compiled by the Oregon Department of Geology and Mineral Industries and includes detailed descriptions, histories and the geology of over 750 metallic mineral mines and prospects in North Eastern Oregon. **8.5" X 11", 310 ppgs. Retail Price: $29.99**

Metal Mines of North West Oregon - Oregon mining historian Kerby Jackson introduces us to a classic work on Oregon's mining history in this important re-issue of Bulletin 14D, otherwise known as the North West Oregon Metal Mines Handbook. Unavailable since 1951, this important publication was originally compiled by the Oregon Department of Geology and Mineral Industries and includes detailed descriptions, histories and the geology of over 250 metallic mineral mines and prospects in North Western Oregon. **8.5" X 11", 182 ppgs. Retail Price: $19.99**

Mines and Prospects of Oregon - Mining historian Kerby Jackson introduces us to a classic mining work by the Oregon Bureau of Mines in this important re-issue of The Handbook of Mines and Prospects of Oregon. Unavailable since 1916, this publication includes important insights into hundreds of gold, silver, copper, coal, limestone and other mines that operated in the State of Oregon around the turn of the 19th Century. Included are not only geological details on early mines throughout Oregon, but also insights into their history, production, locations and in some cases, also included are rare maps of their underground workings. **8.5" X 11", 314 ppgs. Retail Price: $24.99**

Lode Gold of the Klamath Mountains of Northern California and South West Oregon
(See California Mining Books)

Mineral Resources of South West Oregon - Unavailable since 1914, this publication includes important insights into dozens of mines that once operated in South West Oregon, including the famous gold fields of Josephine and Jackson Counties, as well as the Coal Mines of Coos County. Included are not only geological details on early mines throughout South West Oregon, but also insights into their history, production and locations. **8.5" X 11", 154 ppgs. Retail Price: $11.99**

Chromite Mining in The Klamath Mountains of California and Oregon
(See California Mining Books)

Southern Oregon Mineral Wealth - Unavailable since 1904, this rare publication provides a unique snapshot into the mines that were operating in the area at the time. Included are not only geological details on early mines throughout South West Oregon, but also insights into their history, production and locations. Some of the mining areas include Grave Creek, Greenback, Wolf Creek, Jump Off Joe Creek, Granite Hill, Galice, Mount Reuben, Gold Hill, Galls Creek, Kane Creek, Sardine Creek, Birdseye Creek, Evans Creek, Foots Creek, Jacksonville, Ashland, the Applegate River, Waldo, Kerby and the Illinois River, Althouse and Sucker Creek, as well as insights into local copper mining and other topics. **8.5" X 11", 64 ppgs. Retail Price: $8.99**

Geology and Ore Deposits of the Takilma and Waldo Mining Districts - Unavailable since the 1933, this publication was originally compiled by the United States Geological Survey and includes details on gold and copper mining in the Takilma and Waldo Districts of Josephine County, Oregon. The Waldo and Takilma mining districts are most notable for the fact that the earliest large scale mining of placer gold and copper in Oregon took place in these two areas. Included in this report are details about some of the earliest large gold mines in the state such as the Llano de Oro, High Gravel, Cameron, Platerica, Deep Gravel and others, as well as copper mines such as the famous Queen of Bronze mine, the Waldo, Lily and Cowboy mines. In addition to geological examinations, insights are also provided into the production, day to day operations and early histories of these mines, as well as calculations of known mineral reserves in the area. This volume also includes six maps and 20 original illustrations. **8.5" X 11", 74 ppgs. Retail Price: $9.99**

Gold Mines of Oregon - Oregon mining historian Kerby Jackson introduces us to a classic work on Oregon's mining history in this important re-issue of Bulletin 61, otherwise known as "Gold and Silver In Oregon". Unavailable since 1968, this important publication was originally compiled by geologists Howard C. Brooks and Len Ramp of the Oregon Department of Geology and Mineral Industries and includes detailed descriptions, histories and the geology of over 450 gold mines Oregon. Included are notes on the history, geology and gold production statistics of all the major mining areas in Oregon including the Klamath Mountains, the Blue Mountains and the North Cascades. While gold is where you find it, as every miner knows, the path to success is to prospect for gold where it was previously found. 8.5" X 11", 344 ppgs. Retail Price: $24.99

Mines and Mineral Resources of Curry County Oregon - Originally published in 1916, this important publication on Oregon Mining has not been available for nearly a century. Included are rare insights into the history, production and locations of dozens of gold mines in Curry County, Oregon, as well as detailed information on important Oregon mining districts in that area such as those at Agness, Bald Face Creek, Mule Creek, Boulder Creek, China Diggings, Collier Creek, Elk River, Gold Beach, Rock Creek, Sixes River and elsewhere. Particular attention is especially paid to the famous beach gold deposits of this portion of the Oregon Coast. 8.5" X 11", 140 ppgs. Retail Price: $11.99

Chromite Mining in South West Oregon - Originally published in 1961, this important publication on Oregon Mining has not been available for nearly a century. Included are rare insights into the history, production and locations of nearly 300 chromite mines in South Western Oregon. 8.5" X 11", 184 ppgs. Retail Price: $14.99

Mineral Resources of Douglas County Oregon - Originally published in 1972, this important publication on Oregon Mining has not been available for nearly forty years. Included are rare insights into the geology, history, production and locations of numerous gold mines and other mining properties in Douglas County, Oregon. 8.5" X 11", 124 ppgs. Retail Price: $11.99

Mineral Resources of Coos County Oregon - Originally published in 1972, this important publication on Oregon Mining has not been available for nearly forty years. Included are rare insights into the geology, history, production and locations of numerous gold mines and other mining properties in Coos County, Oregon. 8.5" X 11", 100 ppgs. Retail Price: $11.99

Mineral Resources of Lane County Oregon - Originally published in 1938, this important publication on Oregon Mining has not been available for nearly seventy five years. Included are extremely rare insights into the geology and mines of Lane County, Oregon, in particular in the Bohemia, Blue River, Oakridge, Black Butte and Winberry Mining Districts. 8.5" X 11", 82 ppgs. Retail Price: $9.99

Mineral Resources of the Upper Chetco River of Oregon: Including the Kalmiopsis Wilderness - Originally published in 1975, this important publication on Oregon Mining has not been available for nearly forty years. Withdrawn under the 1872 Mining Act since 1984, real insight into the minerals resources and mines of the Upper Chetco River has long been unavailable due to the remoteness of the area. Despite this, the decades of battle between property owners and environmental extremists over the last private mining inholding in the area has continued to pique the interest of those interested in mining and other forms of natural resource use. Gold mining began in the area in the 1850's and has a rich history in this geographic area, even if the facts surrounding it are little known. Included are twenty two rare photographs, as well as insights into the Becca and Morning Mine, the Emmly Mine (also known as Emily Camp), the Frazier Mine, the Golden Dream or Higgins Mine, Hustis Mine, Peck Mine and others. 8.5" X 11", 64 ppgs. Retail Price: $8.99

Gold Dredging in Oregon - Originally published in 1939, this important publication on Oregon Mining has not been available for nearly seventy five years. Included are extremely rare insights into the history and day to day operations of the dragline and bucketline gold dredges that once worked the placer gold fields of South West and North East Oregon in decades gone by. Also included are details into the areas that were worked by gold dredges in Josephine, Jackson, Baker and Grant counties, as well as the economic factors that impacted this mining method. This volume also offers a unique look into the values of river bottom land in relation to both farming and mining, in how farm lands were mined, re-soiled and reclamated after the dredges worked them. Featured are hard to find maps of the gold dredge fields, as well as rare photographs from a bygone era. 8.5" X 11", 86 ppgs. Retail Price: $8.99

Quick Silver Mining in Oregon - Originally published in 1963, this important publication on Oregon Mining has not been available for over fifty years. This publication includes details into the history and production of Elemental Mercury or Quicksilver in the State of Oregon. 8.5" X 11", 238 ppgs. Retail Price: $15.99

Mines of the Greenhorn Mining District of Grant County Oregon - Originally published in 1948, this important publication on Oregon Mining has not been available for over sixty five years. In this publication are rare insights into the mines of the famous Greenhorn Mining District of Grant County, Oregon, especially the famous Morning Mine. Also included are details on the Tempest, Tiger, Bi-Metallic, Windsor, Psyche, Big Johnny, Snow Creek, Banzette and Paramount Mines, as well as prospects in the vicinities in the famous mining areas of Mormon Basin, Vinegar Basin and Desolation Creek. Included are hard to find mine maps and dozens of rare photographs from the bygone era of Grant County's rich mining history. 8.5" X 11", 72 ppgs. Retail Price: $9.99

Geology of the Wallowa Mountains of Oregon: Part I (Volume 1) - Originally published in 1938, this important publication on Oregon Mining has not been available for nearly seventy five years. Included are details on the geology of this unique portion of North Eastern Oregon. This is the first part of a two book series on the area. Accompanying the text are rare photographs and historic maps. 8.5" X 11", 92 ppgs. Retail Price: $9.99

Geology of the Wallowa Mountains of Oregon: Part II (Volume 2) - Originally published in 1938, this important publication on Oregon Mining has not been available for nearly seventy five years. Included are details on the geology of this unique portion of North Eastern Oregon. This is the first part of a two book series on the area. Accompanying the text are rare photographs and historic maps. 8.5" X 11", 94 ppgs. Retail Price: $9.99

Field Identification of Minerals For Oregon Prospectors - Originally published in 1940, this important publication on Oregon Mining has not been available for nearly seventy five years. Included in this volume is an easy system for testing and identifying a wide range of minerals that might be found by prospectors, geologists and rockhounds in the State of Oregon, as well as in other locales. Topics include how to put together your own field testing kit and how to conduct rudimentary tests in the field. This volume is written in a clear and concise way to make it useful even for beginners. 8.5" X 11", 158 ppgs. Retail Price: $14.99

The Bohemia Mining District of Oregon - Originally published in 1900, this important publication on Oregon Mining has not been available for over a century. Included in this volume are important insights into the famous Bohemia Mining District of Oregon, including the histories and locations of important gold mines in the area such as the Ophir Mine, Clarence, Acturas, Peek-a-boo, White Swan, Combination Mine, the Musick Mine, The California, White Ghost, The Mystery, Wall Street, Vesuvius, Story, Lizzie Bullock, Delta, Elsie Dora, Golden Slipper, Broadway, Champion Mine, Knott, Noonday, Helena, White Wings, Riverside and others. Also included are notes on the nearby Blue River Mining District. 8.5" X 11", 58 ppgs. Retail Price: $9.99

The Gold Fields of Eastern Oregon - Unavailable since 1900, this publication was originally compiled by the Baker City Chamber of Commerce Offering important insights into the gold mining history of Eastern Oregon, "The Gold Fields of Eastern Oregon" sheds a rare light on many of the gold mines that were operating at the turn of the 19th Century in Baker County and Grant County in North Eastern Oregon. Some of the areas featured include the Cable Cove District, Baisely-Elhorn, Granite, Red Boy, Bonanza, Susanville, Sparta, Virtue, Vaughn, Sumpter, Burnt River, Rye Valley and other mining districts. Included is basic information on not only many gold mines that are well known to those interested in Eastern Oregon mining history, but also many mines and prospects which have been mostly lost to the passage of time. Accompanying are numerous rare photos 8.5" X 11", 78 ppgs. Retail Price: $10.99

Gold Mining in Eastern Oregon - Originally published in 1938, this important publication on Oregon Mining has not been available for over a century. Included in this volume are important insights into the famous mining districts of Eastern Oregon during the late 1930's. Particular attention is given to those gold mines with milling and concentrating facilities in the Greenhorn, Red Boy, Alamo, Bonanza, Granite, Cable Cove, Cracker Creek, Virtue, Keating, Medical Springs, Sanger, Sparta, Chicken Creek, Mormon Basin, Connor Creek, Cornucopia and the Bull Run Mining Districts. Some of the mines featured include the Ben Harrison, North Pole-Columbia, Highland Maxwell, Baisley-Elkhorn, White Swan, Balm Creek, Twin Baby, Gem of Sparta, New Deal, Gleason, Gifford-Johnson, Cornucopia, Record, Bull Run, Orion and others. Of particular interest are the mill flow sheets and descriptions of milling operations of these mines. 8.5" X 11", 68 ppgs. Retail Price: $8.99

The Gold Belt of the Blue Mountains of Oregon - Originally published in 1901, this important publication on Oregon Mining has not been available for over a century. Included in this volume are rare insights into the gold deposits of the Blue Mountains of North East Oregon, including the history of their early discovery and early production. Extensive details are offered on this important mining area's mineralogy and economic geology, as well as insights into nearby gold placers, silver deposits and copper deposits. Featured are the Elkhorn and Rock Creek mining districts, the Pocahontas district, Auburn and Minersville districts, Sumpter and Cracker Creek, Cable Cove, the Camp Carson district, Granite, Alamo, Greenhorn, Robinsonville, the Upper Burnt River Valley and Bonanza districts, Susanville, Quartzburg, Canyon Creek, Virtue, the Copper Butte district, the North Powder River, Sparta, Eagle Creek, Cornucopia, Pine Creek, Lower Powder River, the Upper Snake River Canyon, Rye Valley, Lower Burnt River Valley, Mormon Basin, the Malheur and Clarks Creek districts, Sutton Creek and others. Of particular interest are important details on numerous gold mines and prospects in these mining districts, including their locations, histories, geology and other important information, as well as information on silver, copper and fire opal deposits. 8.5" X 11", 250 ppgs. Retail Price: $24.99

Mining in the Cascades Range of Oregon - Originally published in 1938, this important publication on Oregon Mining has not been available for over seventy five years. Included in this volume are rare insights into the gold mines and other types of metal mines in the Cascades Mountain Range of Oregon. Some of the important mining areas covered include the famous Bohemia Mining District, the North Santiam Mining District, Quartzville Mining District, Blue River Mining District, Fall Creek Mining District, Oakridge District, Zinc District, Buzzard-Al Sarena District, Grand Cove, Climax District and Barron Mining District. Of particular interest are important details on over 100 mines and prospects in these mining districts, including their locations, histories, geology and other important information. **8.5″ X 11″, 170 ppgs. Retail Price: $14.99**

Beach Gold Placers of the Oregon Coast - Originally published in 1934, this important publication on Oregon Mining has not been available for over 80 years. Included in this volume are rare insights into the beach gold deposits of the State of Oregon, including their locations, occurance, composition and geology. Of particular interest is information on placer platinum in Oregon's rich beach deposits. Also included are the locations and other information on some famous Oregon beach mines, including the Pioneer, Eagle, Chickamin, Iowa and beach placer mines north of the mouth of the Rogue River. **8.5″ X 11″, 60 ppgs. Retail Price: $8.99**

Mineralogical Composition of the Sands of the Oregon Coast: From Coos Bay to the Columbia - Published in 1945, he text features hard to find information on the composition of the gold bearing black sands of the South West Oregon Coast, offering a unique insight to prospectors in search of Oregon's legendary beach gold. **104 ppgs, $9.99**

Manganese Mining in Oregon - First released in 1942 and now out of print, this special reprint edition of "Manganese in Oregon" was originally published by the Oregon Department of Geology and Mineral Industries. The text features hard to find information on the mining of Manganese in Oregon, including details and maps of Oregon manganese mines and prospects. **108 ppgs, 9.99**

Medford Oregon As A Mining Center - Written in 1912, this hard to find publication includes valuable insights into the mining history of South West Oregon. This small book contains interesting information on the gold, copper and mining industry in Southern Oregon as it existed just prior to World War One, shedding light on some of the important mines in the area. Included are rare photographs and vintage advertising of the day. **80 ppgs, 9.99**

Mineral Resources of Curry County Oregon - First released in 1977 and now out of print, this special reprint edition of "Geology, Mineral Resources and Rock Materials of Curry County, Oregon" was originally published in cooperation of Curry County, Oregon and the Oregon Department of Geology and Mineral Industries. The text features hard to find information on not only the mining of gold and other metals in Curry County, but also aggregate mining in the area. **102 ppgs, 11.99**

Origin of the Gold Bearing Black Sands of the Coast of South West Oregon - First released in 1943 and now out of print, this special reprint edition of "The Origin of the Black Sands of the South West Oregon Coast" was originally published by the Oregon Department of Geology and Mineral Industries. The text features hard to find information on the origin of the gold bearing black sands of the South West Oregon Coast, offering a unique insight to prospectors in search of Oregon's legendary beach gold. **52 ppgs, 8.99**

South West Oregon Mining - Leading mining historian Kerby Jackson introduces us to six classic small mining publications on the Gold Mining Industry in Southern Oregon. This small book consists of a compilation of USGS J.S. Diller's "Mines of the Riddles Quadrangle", "The Rogue River Valley Coal Fields" and "Mineral Resources of the Grants Pass Quadrangle", the Grants Pass Commercial Club's rare publication "Mining in Josephine County, Oregon" and the USGS publication "The Distribution of Placer Gold in the Sixes River, South West Oregon". Also included is F.W. Libbey's legendary article on the Southern Oregon Mining Industry, "Lest We Forget", which appeared in the publication of the Oregon State Department of Geology and Mineral Industries in the early 1960's. This compilation offers a unique perspective on mining in South West Oregon and includes considerable information on mines in Josephine, Jackson and Coos Counties. **142 ppgs, 14.99**

Geology and Mineral Resources of the Gasquet Quadrangle of California-Oregon - First published in 1953, it has been unavailable for over a century and sheds important light on the geological features and mineral resources of this portion of Northern California and Southern Oregon. **80 ppgs, 9.99**

Idaho Mining Books

Gold in Idaho - Unavailable since the 1940's, this publication was originally compiled by the Idaho Bureau of Mines and includes details on gold mining in Idaho. Included is not only raw data on gold production in Idaho, but also valuable insight into where gold may be found in Idaho, as well as practical information on the gold bearing rocks and other geological features that will assist those looking for placer and lode gold in the State of Idaho. This volume also includes thirteen gold maps that greatly enhance the practical usability of the information contained in this small book detailing where to find gold in Idaho. **8.5" X 11", 72 ppgs. Retail Price: $9.99**

Geology of the Couer D'Alene Mining District of Idaho - Unavailable since 1961, this publication was originally compiled by the Idaho Bureau of Mines and Geology and includes details on the mining of gold, silver and other minerals in the famous Coeur D'Alene Mining District in Northern Idaho. Included are details on the early history of the Coeur D'Alene Mining District, local tectonic settings, ore deposit features, information on the mineral belts of the Osburn Fault, as well as detailed information on the famous Bunker Hill Mine, the Dayrock Mine, Galena Mine, Lucky Friday Mine and the infamous Sunshine Mine. This volume also includes sixteen hard to find maps. **8.5" X 11", 70 ppgs. Retail Price: $9.99**

The Gold Camps and Silver Cities of Idaho - Originally published in 1963, this important publication on Idaho Mining has not been available for nearly fifty years. Included are rare insights into the history of Idaho's Gold Rush, as well as the mad craze for silver in the Idaho Panhandle. Documented in fine detail are the early mining excitements at Boise Basin, at South Boise, in the Owyhees, at Deadwood, Long Valley, Stanley Basin and Robinson Bar, at Atlanta, on the famous Boise River, Volcano, Little Smokey, Banner, Boise Ridge, Hailey, Leesburg, Lemhi, Pearl, at South Mountain, Shoup and Ulysses, Yellow Jacket and Loon Creek. The story follows with the appearance of Chinese miners at the new mining camps on the Snake River, Black Pine, Yankee Fork, Bay Horse, Clayton, Heath, Seven Devils, Gibbonsville, Vienna and Sawtooth City. Also included are special sections on the Idaho Lead and Silver mines of the late 1800's, as well as the mining discoveries of the early 1900's that paved the way for Idaho's modern mining and mineral industry. Lavishly illustrated with rare historic photos, this volume provides a one of a kind documentary into Idaho's mining history that is sure to be enjoyed by not only modern miners and prospectors who still scour the hills in search of nature's treasures, but also those enjoy history and tromping through overgrown ghost towns and long abandoned mining camps. **8.5" X 11", 186 ppgs. Retail Price: $14.99**

Ore Deposits and Mining in North Western Custer County Idaho - Unavailable since 1913, this important publication was originally published by the Us Department of the Interior and has been unavailable for a century. Included are fine details on the geology, geography, gold placers and gold and silver bearing quartz veins of the mining region of North West Custer County, Idaho. Of particular interest is a rare look at the mines and prospects of the region, including those such as the Ramshorn Mine, SkyLark, Riverview, Excelsior, Beardsley, Pacific, Hoosier, Silver Brick, Forest Rose and dozens of others in the Bay Horse Mining District. Also covered are the mines of the Yankee Fork District such as the Lucky Boy, Badger, Black, Enterprise, Charles Dickens, Morrison, Golden Sunbeam, Montana, Golden Gate and others, as well as those in the Loon Mining District. **8.5" X 11", 126 ppgs. Retail Price: $12.99**

Gold Rush To Idaho - Unavailable since 1963, this important publication was originally published by the Idaho Bureau of Mines and has been unavailable for 50 years. "Gold Rush To Idaho" revisits the earliest years of the discovery of gold in Idaho Territory and introduces us to the conditions that the pioneer gold seekers met when they blazed a trail through the wilderness of Idaho's mountains and discovered the precious yellow metal at Oro Fino and Pierce. Subsequent rushes followed at places like Elk City, Newsome, Clearwater Station, Florence, Warrens and elsewhere. Of particular interest is a rare look at the hardships that the first miners in Idaho met with during their day to day existences and their attempts to bring law and order to their mining camps. **8.5" X 11", 88 ppgs. Retail Price: $9.99**

The Geology and Mines of Northern Idaho and North Western Montana - Unavailable since 1909, this important publication was originally published by the Us Department of the Interior and has been unavailable for a century. Included are fine details on the geology and geography of the mining regions of Northern Idaho and North Western Montana. Of particular interest is a rare look at the mines and prospects of the region, including those in the Pine Creek Mining District, Lake Pend Oreille district, Troy Mining District, Sylvanite District, Cabinet Mining District, Prospect Mining District and the Missoula Valley. Some of the mines featured include the Iron Mountain, Silver Butte, Snowshoe, Grouse Mountain Mine and others. **8.5" X 11", 142 ppgs. Retail Price: $12.99**

Mining in the Alturas Quadrangle of Blaine County Idaho - Unavailable since 1922, this important publication was originally published by the Idaho Bureau of Mines and has been unavailable for ninety years. Topics include the geology, rock formations and the formation of ore deposits in this important mining area of Idaho. Of particular focus is information on the local geology, quartz veins and ore deposits of this portion of Idaho. Included are hard to find details, including the descriptions and locations of numerous gold and silver mines in the area including the Silver King, Pilgrim, Columbia, Lone Jack, Sunbeam, Pride of the West, Lucky Boy, Scotia, Atlanta, Beaver-Bidwell and others mines and prospects. **8.5" X 11", 56 ppgs. Retail Price: $8.99**

Mining in Lemhi County Idaho - Originally published in 1913, this important book on Idaho Mining has not been available to miners for over a century. Included are rare insights into hundreds of gold, silver, copper and other mines in this famous Idaho mining area. Details include the locations, geology, history, production and other facts of the mines of this region, not only gold and silver hardrock mines, but also gold placer mines, lead-silver deposits, copper mines, cobalt-nickel deposits, tungsten and tin mines . It is lavishly illustrated with hard to find photos of the period and rare mining maps. Some of the vicinities featured include the Nicholia Mining District, Spring Mountain District, Texas District, Blue Wing District, Junction District, McDevitt District, Pratt Creek, Eldorado District, Kirtley Creek, Carmen Creek, Gibbonsville, Indian Creek, Mineral Hill District, Mackinaw, Eureka District, Blackbird District, YellowJacket District, Gravel Range District, Junction District, Parker Mountain and other mining districts. **8.5" X 11", 226 ppgs. Retail Price: $19.99**

Mining in Shoshone County Idaho - First published in 1923, it has been unavailable for over a century and sheds important light on the mining history of Shoshone County, Idaho. Some of the topics include the history of mining in Shoshone County, a look at the local geology and ore characteristics of lead-silver deposits, zinc deposits, copper, antimony, gold and other minerals. Also included are insights into the history, production, characteristics and locations of numerous mines in the area. 198 ppgs, 15.99

Utah Mining Books

Fluorite in Utah - Unavailable since 1954, this publication was originally compiled by the USGS, State of Utah and U.S. Atomic Energy Commission and details the mining of fluorspar, also known as fluorite in the State of Utah. Included are details on the geology and history of fluorspar (fluorite) mining in Utah, including details on where this unique gem mineral may be found in the State of Utah. **8.5" X 11", 60 ppgs. Retail Price: $8.99**

The Gold Hill Mining District of Utah - First published in 1935, it has been unavailable since those days and sheds important light on the mines, history and geology of Utah's Gold Hill Mining District. Included are rare insights into this important mining area, including the locations, histories and details of numerous mines. This volume is well illustrated with geological diagrams, as well as hard to find maps of some of the most important mines in this district. 202 ppgs., 19.99

The Mines, Miners and Minerals of Utah - First published in 1896, it has been unavailable since those days and sheds important light on the early mines and miners of Pioneer Utah, as well as the minerals which they won from the earth by laborious hard physical labor and sheer determination. Included are rare insights into the early mining history of Utah, as well details on hundreds of gold, silver and copper mines. 376 ppgs., 24.99

California Mining Books

The Tertiary Gravels of the Sierra Nevada of California - Mining historian Kerby Jackson introduces us to a classic mining work by Waldemar Lindgren in this important re-issue of The Tertiary Gravels of the Sierra Nevada of California. Unavailable since 1911, this publication includes details on the gold bearing ancient river channels of the famous Sierra Nevada region of California. **8.5" X 11", 282 ppgs. Retail Price: $19.99**

The Mother Lode Mining Region of California - Unavailable since 1900, this publication includes details on the gold mines of California's famous Mother Lode gold mining area. Included are details on the geology, history and important gold mines of the region, as well as insights into historic mining methods, mine timbering, mining machinery, mining bell signals and other details on how these mines operated. Also included are insights into the gold mines of the California Mother Lode that were in operation during the first sixty years of California's mining history. **8.5" X 11", 176 ppgs. Retail Price: $14.99**

Lode Gold of the Klamath Mountains of Northern California and South West Oregon - Unavailable since 1971, this publication was originally compiled by Preston E. Hotz and includes details on the lode mining districts of Oregon and California's Klamath Mountains. Included are details on the geology, history and important lode mines of the French Gulch, Deadwood, Whiskeytown, Shasta, Redding, Muletown, South Fork, Old Diggings, Dog Creek (Delta), Bully Choop (Indian Creek), Harrison Gulch, Hayfork, Minersville, Trinity Center, Canyon Creek, East Fork, New River, Denny, Liberty (Black Bear), Cecilville, Callahan, Yreka, Fort Jones and Happy Camp mining districts in California, as well as the Ashland, Rogue River, Applegate, Illinois River, Takilma, Greenback, Galice, Silver Peak, Myrtle Creek and Mule Creek districts of South Western Oregon. Also included are insights into the mineralization and other characteristics of this important mining region. **8.5" X 11", 100 ppgs. Retail Price: $10.99**

Mines and Mineral Resources of Shasta County, Siskiyou County, Trinity County: California - Unavailable since 1915, this publication was originally compiled by the California State Mining Bureau and includes details on the gold mines of this area of Northern California. Also included are insights into the mineralization and other characteristics of this important mining region, as well as the location of historic gold mines. **8.5" X 11", 204 ppgs. Retail Price: $19.99**

Geology of the Yreka Quadrangle, Siskiyou County, California - Unavailable since 1977, this publication was originally compiled by Preston E. Hotz and includes details on the geology of the Yreka Quadrangle of Siskiyou County, California. Also included are insights into the mineralization and other characteristics of this important mining region. 8.5" X 11", 78 ppgs. **Retail Price: $7.99**

Mines of San Diego and Imperial Counties, California - Originally published in 1914, this important publication on California Mining has not been available for a century. This publication includes important information on the early gold mines of San Diego and Imperial County, which were some of the first gold fields mined in California by early Spanish and Mexican miners before the 49ers came on the scene. Included are not only details on early mining methods in the area, production statistics and geological information, but also the location of the early gold mines that helped make California "The Golden State". Also included are details on the mining of other minerals such as silver, lead, zinc, manganese, tungsten, vanadium, asbestos, barite, borax, cement, clay, dolomite, fluospar, gem stones, graphite, marble, salines, petroleum, stronium, talc and others. 8.5" X 11", 116 ppgs. **Retail Price: $12.99**

Mines of Sierra County, California - Unavailable since 1920, this publication was originally compiled by the California State Mining Bureau and includes details on the gold mines of Sierra County, California. Also included are insights into the mineralization and other characteristics of this important mining region, as well as the location of historic gold mines. 8.5" X 11", 156 ppgs. **Retail Price: $19.99**

Mines of Plumas County, California - Unavailable since 1918, this publication was originally compiled by the California State Mining Bureau and includes details on the gold mines of Plumas County, California. Also included are insights into the mineralization and other characteristics of this important mining region, as well as the location of historic gold mines. 8.5" X 11", 200 ppgs. **Retail Price: $19.99**

Mines of El Dorado, Placer, Sacramento and Yuba Counties, California - Originally published in 1917, this important publication on California Mining has not been available for nearly a century. This publication includes important information on the early gold mines of El Dorado County, Placer County, Sacramento County and Yuba County, which were some of the first gold fields mined by the Forty-Niners during the California Gold Rush. Included are not only details on early mining methods in the area, production statistics and geological information, but also the location of the early gold mines that helped make California "The Golden State". Also included are insights into the early mining of chrome, copper and other minerals in this important mining area. 8.5" X 11", 204 ppgs. **Retail Price: $19.99**

Mines of Los Angeles, Orange and Riverside Counties, California - Originally published in 1917, this important publication on California Mining has not been available for nearly a century. This publication includes important information on the early gold mines of Los Angeles County, Orange County and Riverside County, which were some of the first gold fields mined in California by early Spanish and Mexican miners before the 49ers came on the scene. Included are not only details on early mining methods in the area, production statistics and geological information, but also the location of the early gold mines that helped make California "The Golden State". 8.5" X 11", 146 ppgs. **Retail Price: $12.99**

Mines of San Bernadino and Tulare Counties, California - Originally published in 1917, this important publication on California Mining has not been available for nearly a century. This publication includes important information on the early gold mines of San Bernadino and Tulare County, which were some of the first gold fields mined in California by early Spanish and Mexican miners before the 49ers came on the scene. Included are not only details on early mining methods in the area, production statistics and geological information, but also the location of the early gold mines that helped make California "The Golden State". Also included are details on the mining of other minerals such as copper, iron, lead, zinc, manganese, tungsten, vanadium, asbestos, barite, borax, cement, clay, dolomite, fluospar, gem stones, graphite, marble, salines, petroleum, stronium, talc and others. 8.5" X 11", 200 ppgs. **Retail Price: $19.99**

Chromite Mining in The Klamath Mountains of California and Oregon - Unavailable since 1919, this publication was originally compiled by J.S. Diller of the United States Department of Geological Survey and includes details on the chromite mines of this area of Northern California and Southern Oregon. Also included are insights into the mineralization and other characteristics of this important mining region, as well as the location of historic mines. Also included are insights into chromite mining in Eastern Oregon and Montana. 8.5" X 11", 98 ppgs. **Retail Price: $9.99**

Mines and Mining in Amador, Calaveras and Tuolumne Counties, California - Unavailable since 1915, this publication was originally compiled by William Tucker and includes details on the mines and mineral resources of this important California mining area. Included are details on the geology, history and important gold mines of the region, as well as insights into other local mineral resources such as asbestos, clay, copper, talc, limestone and others. Also included are insights into the mineralization and other characteristics of this important portion of California's Mother Lode mining region. 8.5" X 11", 198 ppgs. **Retail Price: $14.99**

The Cerro Gordo Mining District of Inyo County California - Unavailable since 1963, this publication was originally compiled by the United States Department of Interior. Included are insights into the mineralization and other characteristics of this important mining region of Southern California. Topics include the mining of gold and silver in this important mining district in Inyo County, California, including details on the history, production and locations of the Cerro Gordo Mine, the Morning Star Mine, Estelle Tunnel, Charles Lease Tunnel, Ignacio, Hart, Crosscut Tunnel, Sunset, Upper Newtown, Newtown, Ella, Perseverance, Newsboy, Belmont and other silver and gold mines in the Cerro Gordo Mining District. This volume also includes important insights into the fossil record, geologic formations, faults and other aspects of economic geology in this California mining district. 8.5" X 11", 104 ppgs. Retail Price: $10.99

Mining in Butte, Lassen, Modoc, Sutter and Tehama Counties of California - Unavailable since 1917, this publication was originally compiled by the United States Department of Interior. Included are insights into the mineralization and other characteristics of this important mining region of California. Topics include the mining of asbestos, chromite, gold, diamonds and manganese in Butte County, the mining of gold and copper in the Hayden Hill and Diamond Mountain mining districts of Lassen County, the mining of coal, salt, copper and gold in the High Grade and Winters mining districts of Modoc County, gold mining in Sutter County and the mining of gold, chromite, manganese and copper in Tehama County. This volume also includes the production records and locations of numerous mines in this important mining region. 8.5" X 11", 114 ppgs. Retail Price: $11.99

Mines of Trinity County California - Originally published in 1965, this important publication on California Mining has not been available for nearly fifty years. This publication includes important information on mines and mining in Trinity County, California, as well insights into the mineralization and geology of this important mining area in Northern California. Included are extensive details on hardrock and placer gold mines and prospects, including charts showing the locations of these historic mines.. 8.5" X 11", 144 ppgs. Retail Price: $12.99

Mines of Kern County California - Originally published in 1962, this important publication on California Mining has not been available for nearly fifty years. This publication includes important information on mines and mining in Kern County, California, as well insights into the mineralization and geology of this important mining area in California. Included are extensive details on hardrock and placer gold mines and prospects, including charts showing the locations of these historic mines. 8.5" X 11", 398 ppgs. Retail Price: $24.99

Mines of Calaveras County California - Originally published in 1962, this important publication on California Mining has not been available for nearly fifty years. This publication includes important information on mines and mining in Calaveras County, California, as well insights into the mineralization and geology of this important mining area in Northern California. Included are extensive details on hardrock and placer gold mines and prospects, including charts showing the locations of these historic mines. 8.5" X 11", 236 ppgs. Retail Price: $19.99

Lode Gold Mining in Grass Valley California - Unavailable since 1940, this publication was originally compiled by the United States Department of Interior. Included are insights into the gold mineralization and other characteristics of this important mining region of Nevada County, California. This volume also includes important insights into the geologic formations, faults and other aspects of economic geology in this California mining district. Of particular interest are the fine details on many hardrock gold mines in the area, including their locations, histories, development and mineralization. Some of the mines featured include the Gold Hill Mine, Massachusetts Hill, Boundary, Peabody, Golden Center, North Star, Omaha, Lone Jack, Homeward Bound, Hartery, Wisconsin, Allison Ranch, Phoenix, Kate Hayes, W.Y.O.D., Empire, Rich Hill, Daisy Hill, Orleans, Sultana, Centennial, Conlin, Ben Franklin, Crown Point and many others. 8.5" X 11", 148 ppgs. Retail Price: $12.99

Lode Mining in the Alleghany District of Sierra County California - Unavailable since 1913, this publication was originally compiled by the United States Department of Interior. Included are insights into the mineralization and other characteristics of this important mining region of Sierra County. Included are details on the history, production and locations of numerous hardrock gold mines in this famous California area, including the Tightner Mine, Minnie D., Osceola, Eldorado, Twenty One, Sherman, Kenton, Oriental, Rainbow, Plumbago, Irelan, Gold Canyon, North Fork, Federal, Kate Hardy and others. This volume also includes important insights into the fossil record, geologic formations, faults and other aspects of economic geology in this California mining district. 8.5" X 11", 48 ppgs. Retail Price: $7.99

Six Months In The Gold Mines During The California Gold Rush - Unavailable since 1850, this important work is a first hand account of one "49'ers" personal experience during the great California Gold Rush, shedding important light on one of the most exciting periods in the history of not only California, but also the world. Compiled from journals written between 1847 and 1849 by E. Gould Buffum, a native of New York, "Six Months In The Gold Mines During The California Gold Rush" offers a rare look into the day to day lives of the people who came to California to work in her gold mines when the state was still a great frontier. 8.5" X 11", 290 ppgs. Retail Price: $19.99

<u>**Quartz Mines of the Grass Valley Mining District of California**</u> - Unavailable since 1867, this important publication has not been available since those days. This rare publication offers a short dissertation on the early hardrock mines in this important mining district in the California Mother Lode region between the 1850's and 1860's. Also included are hard to find details on the mineralization and locations of these mines, as well as how they were operated in those day. **8.5" X 11", 44 ppgs. Retail Price: $8.99**

<u>**Gold Rush on the Feather River**</u> - First published in 1924, this short publication by G.C. Mansfield sheds important light on the early history of gold mining on the Feather River. Included are rare insights into the first decade of gold mining and the early mining camps of the Feather River during the 1850's. 64 ppgs., 9.99

<u>**The Bodie Mining District of California**</u> - First published in 1986, it has been unavailable since those days and sheds important light on this famous mining area. Included are the history, characteristics and locations of numerous old mines around the ghost town of Bodie. 64 ppgs, 8.99

<u>**Geology and Mineral Resources of the Gasquet Quadrangle of California-Oregon**</u> - First published in 1953, it has been unavailable for over a century and sheds important light on the geological features and mineral resources of this portion of Northern California and Southern Oregon. 80 ppgs, 9.99

Alaska Mining Books

<u>**Ore Deposits of the Willow Creek Mining District, Alaska**</u> - Unavailable since 1954, this hard to find publication includes valuable insights into the Willow Creek Mining District near Hatcher Pass in Alaska. The publication includes insights into the history, geology and locations of the well known mines in the area, including the Gold Cord, Independence, Fern, Mabel, Lonesome, Snowbird, Schroff-O'Neil, High Grade, Marion Twin, Thorpe, Webfoot, Kelly-Willow, Lane, Holland and others. **8.5" X 11", 96 ppgs. Retail Price: $9.99**

<u>**The Juneau Gold Belt of Alaska**</u> - Unavailable since 1906, this hard to find publication includes valuable insights into the gold mines around Juneau, Alaska. The publication includes important details into the history, geology and locations of the well known gold mines and prospects in the area, including those around Windham Bay, Holkham Bay, Port Snettisham, on Grindstone and Rhine Creeks, Gold Creek, Douglas Island, Salmon Creek, Lemon Creek, Nugget Creek, from the Mendenhall River to Berners Bay, McGinnis Creek, Montana Creek, Peterson Creek, Windfall Creek, the Eagle River, Yankee Basin, Yankee Curve, Kowee Creek and elsewhere. Not only are gold placer mines included, but also hardrock gold mines. **8.5" X 11", 224 ppgs. Retail Price: $19.99**

<u>**Mining in the Jumbo Basin of Alaska**</u> - Unavailable since 1953, this hard to find publication includes valuable insights into the mines and geology of the Jumbo Basin. The publication includes important details into the history, geology and locations of the well known gold mines and prospects in the famous Jumbo Basin Mining Region of Alaska. 72 ppgs, 9.99

<u>**The Rampart Placer Gold Region of Alaska**</u> - Unavailable since 1906, this hard to find publication includes valuable insights into the placer gold mines of the Rampart Mining Region. The publication includes important details into the history, geology and locations of the well known gold mines and prospects in the famous Rampart Mining Region of Alaska. 78 ppgs, 10.99

Arizona Mining Books

<u>**Mines and Mining in Northern Yuma County Arizona**</u> - Originally published in 1911, this important publication on Arizona Mining has not been available for over a hundred years. Included are rare insights into the gold, silver, copper and quicksilver mines of Yuma County, Arizona together with hard to find maps and photographs. Some of the mines and mining districts featured include the Planet Copper Mine, Mineral Hill, the Clara Consolidated Mine, Viati Mine, Copper Basin prospect, Bowman Mine, Quartz King, Billy Mack, Carnation, the Wardwell and Osbourne, Valensuella Copper, the Mariquita, Colonial Mine, the French American, the New York-Plomosa, Guadalupe, Lead Camp, Mudersbach Copper Camp, Yellow Bird, the Arizona Northern (Salome Strike), Bonanza (Harqua Hala), Golden Eagle, Hercules, Socorro and others. **8.5" X 11", 144 ppgs. Retail Price: $11.99**

<u>**The Aravaipa and Stanley Mining Districts of Graham County Arizona**</u> - Originally published in 1925, this important publication on Arizona Mining has not been available for nearly ninety years. Included are rare insights into the gold and silver mines of these two important mining districts, together with hard to find maps. **8.5" X 11", 140 ppgs. Retail Price: $11.99**

Gold in the Gold Basin and Lost Basin Mining Districts of Mohave County, Arizona - This volume contains rare insights into the geology and gold mineralization of the Gold Basin and Lost Basin Mining Districts of Mohave County, Arizona that will be of benefit to miners and prospectors. Also included is a significant body of information on the gold mines and prospects of this portion of Arizona. This volume is lavishly illustrated with rare photos and mining maps. 8.5" X 11", 188 ppgs. Retail Price: $19.99

Mines of the Jerome and Bradshaw Mountains of Arizona - This important publication on Arizona Mining has not been available for ninety years. This volume contains rare insights into the geology and ore deposits of the Jerome and Bradshaw Mountains of Arizona that will be of benefit to miners and prospectors who work those areas. Included is a significant body of information on the mines and prospects of the Verde, Black Hills, Cherry Creek, Prescott, Walker, Groom Creek, Hassayampa, Bigbug, Turkey Creek, Agua Fria, Black Canyon, Peck, Tiger, Pine Grove, Bradshaw, Tintop, Humbug and Castle Creek Mining Districts. This volume is lavishly illustrated with rare photos and mining maps. 8.5" X 11", 218 ppgs. Retail Price: $19.99

The Ajo Mining District of Pima County Arizona - This important publication on Arizona Mining has not been available for nearly seventy years. This volume contains rare insights into the geology and mineralization of the Ajo Mining District in Pima County, Arizona and in particular the famous New Cornelia Mine. 8.5" X 11", 126 ppgs. Retail Price: $11.99

Mining in the Santa Rita and Patagonia Mountains of Arizona - Originally published in 1915, this important publication on Arizona Mining has not been available for nearly a century. Included are rare insights into hundreds of gold, silver, copper and other mines in this famous Arizona mining area. Details include the locations, geology, history, production and other facts of the mines of this region. 8.5" X 11", 394 ppgs. Retail Price: $24.99

Mining in the Bisbee Quadrangle of Arizona - Originally published in 1906, this important publication on Arizona Mining has not been available for nearly a century. Included are rare insights into hundreds of gold, silver, copper and other mines in this famous Arizona mining area. Details include the locations, geology, history, production and other facts of the mines of this important mining region. 8.5" X 11", 188 ppgs. Retail Price: $14.99

Placer Gold Mining in Arizona - Unavailable since 1922, this hard to find publication includes valuable insights into the placer gold mines of the Arizona. Originally released as "Placer Gold of Arizona", despite its small size, this publication includes important details into the history, geology and locations of the well known placer gold mines and prospects in the State of Arizona. 48 ppgs, 8.99

Gold and Copper Mining near Payson, Arizona - Written in 1915, this hard to find publication includes valuable insights into the gold and copper mining industry of Arizona. Highlighted here are the gold and copper mines near Payson, Arizona. 68 ppgs, 8.99

Lode Gold Mining in Arizona - Unavailable since 1934, this hard to find publication, originally released as "Arizona Lode Gold Mines and Gold Mining" includes valuable insights into the gold mining industry of Arizona. Included are valuable insights into over 150 hardrock gold mines in over 30 different mining districts in Arizona. 278 ppgs, 21.99

Mining in the Dragoon Quadrangle of Cochise County, Arizona - Unavailable since 1964, this hard to find publication includes valuable insights into the mines of the Dragoon Quadrangle Mining Region. The publication includes important details into the history, geology and locations of the well known mines and prospects in this famous mining region of Arizona. 224 ppgs., 19.99

Directory of Operating Mines in Arizona in 1915 - Unavailable since 1916, this hard to find publication includes valuable insights into the mines of Arizona. This small publication includes a complete list of the mines that were operating in the State of Arizona during 1915 and includes details such as general location, owners and some basic facts about each mining operation. 52 ppgs. 8.99

Arizona Ore Deposits - Unavailable since 1938, this hard to find publication includes valuable insights into some ore deposits of Arizona. Included are valuable insights into the formation and characteristics of valuable ore deposits in the Jerome, Miami, Inspiration, Clifton, Morenci, Ray, Ajo, Eureka, Tombstone and Magma mining districts. Included are details into some of the major gold, silver and copper mines of these important Arizona mining areas. 160 ppgs, 14.99

Montana Mining Books

A History of Butte Montana: The World's Greatest Mining Camp - First published in 1900 by H.C. Freeman, this important publication sheds a bright light on one of the most important mining areas in the history of The West. Together with his insights, as well as rare photographs of the periods, Harry Freeman describes Butte and its vicinity from its early beginnings, right up to its flush years when copper flowed from its mines like a river. At the time of publication, Butte, Montana was known worldwide as "The Richest Mining Spot On Earth" and produced not only vast amounts of copper, but also silver, gold and other metals from its mines. Freeman illustrates, with great detail, the most important mines in the vicinity of Butte, providing rare details on their owners, their history and most importantly, how the mines operated and how their treasures were extracted. Of particular interest are the dozens of rare photographs that depict mines such as the famous Anaconda, the Silver Bow, the Smoke House, Moose, Paulin, Buffalo, Little Minah, the Mountain Consolidated, West Greyrock, Cora, the Green Mountain, Diamond, Bell, Parnell, the Neversweat, Nipper, Original and many others.   8.5" X 11", 142 ppgs. Retail Price: $12.99

The Butte Mining District of Montana - This important publication on Montana Mining has not been available for over a century. Included are rare insights into the gold, copper and silver mines of Butte, Montana together with hard to find maps and photographs. Some of the topics include the early history of gold, silver and copper mining in the Butte area, insight into the geology of its mining areas, the local distribution of gold, silver and copper ores, as well their composition and how to identify them. Also included are detailed facts about the mines in the Butte Mining District, including the famous Anaconda Mine, Gagnon, Parrot, Blue Vein, Moscow, Poulin, Stella, Buffalo, Green Mountain, Wake Up Jim, the Diamond-Bell Group, Mountain Consolidated, East Greyrock, West Greyrock, Snowball, Corra, Speculator, Adirondack, Miners Union, the Jessie-Edith May Group, Otisco, Iduna, Colorado, Lizzie, Cambers, Anderson, Hesperus, Preferencia and dozens of others. 8.5" X 11", 298 ppgs. Retail Price: $24.99

Mines of the Helena Mining Region of Montana - This important publication on Montana Mining has not been available for over a century. Included are rare insights into the gold, copper and silver mines of the vicinity of Helena, Montana, including the Marysville Mining District, Elliston Mining District, Rimini Mining District, Helena Mining District, Clancy Mining District, Wickes Mining District, Boulder and Basin Mining Districts and the Elkhorn Mining District. Some of the topics include the early history of gold, silver and copper mining in the Helena area, insight into the geology of its mining areas, the local distribution of gold, silver and copper ores, as well their composition and how to identify them. Also included are detailed facts, history, geology and locations of over one hundred gold, silver and copper mines in the area . 8.5" X 11", 162 ppgs, Retail Price: $14.99

Mines and Geology of the Garnet Range of Montana - This important publication on Montana Mining has not been available for over a century. Included are rare insights into the gold, copper and silver mines of the vicinity of this important mining area of Montana. Some of the topics include the early history of gold, silver and copper mining in the Garnet Mountains, insight into the geology of its mining areas, the local distribution of gold, silver and copper ores, as well their composition and how to identify them. Also included are detailed facts, history, geology and locations of numerous gold, silver and copper mines in the area . 8.5" X 11", 100 ppgs, Retail Price: $11.99

Mines and Geology of the Philipsburg Quadrangle of Montana - This important publication on Montana Mining has not been available for over a century. Included are rare insights into the gold, copper and silver mines of the vicinity of this important mining area of Montana. Some of the topics include the early history of gold, silver and copper mining in the Philipsburg Quadrangle, insight into the geology of its mining areas, the local distribution of gold, silver and copper ores, as well their composition and how to identify them. Also included are detailed facts, history, geology and locations of over one hundred gold, silver and copper mines in the area 8.5" X 11", 290 ppgs, Retail Price: $24.99

Geology of the Marysville Mining District of Montana - Included are rare insights into the mining geology of the Marysville Mining District. Some of the topics include the early history of gold, silver and copper mining in the area, insight into the geology of its mining areas, the local distribution of gold, silver and copper ores, as well their composition and how to identify them. Also included are detailed facts, history, geology and locations of gold, silver and copper mines in the area 8.5" X 11", 198 ppgs, Retail Price: $19.99

The Geology and Mines of Northern Idaho and North Western Montana- See listing under Idaho.

The History of Gold Dredging in Montana - Unavailable since 1916, this important publication was originally published by the Us Bureau of Mines and has been unavailable for a century. A century and more ago, giant dredging machines dug in Montana's rivers and creeks in search of illusive golden riches. First appearing in California in the 1850's, gold dredges finally reached their peak of development in Siberia and New Zealand before becoming popular again in the United States. This book offers a unique historical perspective on the gold dredges that once operated in Montana. This book on Montana mining history is lavishly illustrated with dozens of rare historic photos gold dredges that once operated in Montana, as well as hard to locate plans on how these dredges were designed. 120 ppgs., 11.99

Nevada Mining Books

The Bull Frog Mining District of Nevada - Unavailable since 1910, this publication was originally compiled by the United States Department of Interior. This volume also includes important insights into the geologic formations, faults and other aspects of economic geology in this Nevada mining district. Of particular interest are the fine details on many mines in the area, including their locations, histories, development and mineralization. Some of the mines featured include the National Bank Mine, Providence, Gibraltor, Tramps, Denver, Original Bullfrog, Gold Bar, Mayflower, Homestake-King and other mines and prospects. **8.5" X 11", 152 ppgs, Retail Price: $14.99**

History of the Comstock Lode - Unavailable since 1876, this publication was originally released by John Wiley & Sons. This volume also includes important insights into the famous Comstock Lode of Nevada that represented the first major silver discovery in the United States. During its spectacular run, the Comstock produced over 192 million ounces of silver and 8.2 million ounces of gold. Not only did the Comstock result in one of the largest mining rushes in history and yield immense fortunes for its owners, but it made important contributions to the development of the State of Nevada, as well as neighboring California. Included here are important details on not only the early development and history of the Comstock, but also rare early insight into its mines, ore and its geology.8.5" X 11", **244 ppgs, Retail Price: $19.99**

The Pioche Mining District of Nevada - First published in 1932, it has been unavailable for over a century and sheds important light on the mining history of Nevada. Some of the topics include the history of mining in this district, as well as the characteristics of its mineral and ore deposits. Also included are insights into the history, production, characteristics and locations of numerous mines in the area. Some of the mines include the Combined Metals, Pioche, Ely Valley, No. 10, Poorman, Wide Awake, Alps, Prince, Virginia Louise, Half Moon, Abe Lincoln, Fairview, Bristol Silver, National, Vesuvius, Inman, Tempest, Hillside, Jackrabbit, Lucky Star, Fortuna, Mendha, Manhattan, Hamburg, Comet, Lyndon and others. 108 ppgs 10.99

The Yerington Mining District of Nevada - First published in 1932, it has been unavailable for over a century and sheds important light on the mining history of Nevada. Some of the topics include the history of mining in this district, as well as the characteristics of its mineral and ore deposits. Also included are insights into the history, production, characteristics and locations of numerous mines in the area. Some of the mines include the Bluestone, Mason Valley, Malachite, McConnell, Greenwood, Western Nevada, Ludwig, Douglas Hill, Casting Copper, Montana-Yerington, Empire, Jim Beatty, Terry and McFarland, Blue Jay and others. 92 ppgs, 10.99

The Genesis of the Ores of Tonopah Nevada - Unavailable since 1918, this hard to find publication includes valuable insights into the gold mines around Tonopah, Nevada. The publication includes important details into the geology of mines in the Tonopah Mining District of Nevada. 90 ppgs, 10.99

Mining Camps of Elko, Lander and Eureka Counties Nevada - Unavailable since 1910, this hard to find publication includes valuable insights into the mining camps of Elko, Lander and Eureka Counties, Nevada. The publication includes important details into the history of mines and mining in these three Nevada counties. 154 ppgs, 12.99

Ore Deposits of the Bullfrog Quadrangle - Unavailable since 1964 and released as "Geology of Bullfrog Quadrangle and Ore Deposits Related to Bullfrog Hills Caldera, Nye County, Nevada and Inyo County, California". The publication includes important details into the geology of mines in the Bullfrog Quadrangle of Nye County, Nevada and Inyo County, California. 52 ppgs, 9.99

Mining in Eureka County Nevada - Unavailable since 1879, this hard to find publication includes valuable insights into the early mining history off Eureka County, Nevada. The publication includes important details into the early history of the mines of Eureka County, as well as their development, production and how their ores were treated. Also included are details on the 1872 Mining Act, as well as the local rules, regulations and customs of the miners in Eureka County.134 ppgs, 12.99

Colorado Mining Books

Ores of The Leadville Mining District - Unavailable since 1926, this publication was originally compiled by the United States Department of Interior. This volume also includes important insights into the ores and mineralization of the Leadville Mining District in Colorado. Topics include historic ore prospecting methods, local geology, insights into ore veins and stockworks, the local trend and distribution of ore channels, reverse faults, shattered rock above replacement ore bodies, mineral enrichment in oxidized and sulphide zones and more. **8.5" X 11", 66 ppgs, Retail Price: $8.99**

Mining in Colorado - Unavailable since 1926, this publication was originally compiled by the United States Department of Interior. This volume also includes important insights into the mining history of Colorado from its early beginnings in the 1850's right up to the mid 1920's. Not only is Colorado's gold mining heritage included, but also its silver, copper, lead and zinc mining industry. Each mining area is treated separately, detailing the development of Colorado's mines on a county by county basis. **8.5" X 11", 284 ppgs, Retail Price: $19.99**

Gold Mining in Gilpin County Colorado - Unavailable since 1876, this publication was originally compiled by the Register Steam Printing House of Central City, Colorado. A rare glimpse at the gold mining history and early mines of Gilpin County, Colorado from their first discovery in the 1850's up to the "flush years" of the mid 1870's. Of particular interest is the history of the discovery of gold in Gilpin County and details about the men who made those first strikes. Special focus is given to the early gold mines and first mining districts of the area, many of which are not detailed in other books on Colorado's gold mining history. **8.5" X 11", 156 ppgs, Retail Price: $12.99**

Mining in the Gold Brick Mining District of Colorado - Important insights into the history of the Gold Brick Mining District, as well as its local geography and economic geology. Also included are the histories and locations of historic mines in this important Colorado Mining District, including the Cortland, Carter, Raymond, Gold Links, Sacramento, Bassick, Sandy Hook, Chronicle, Grand Prize, Chloride, Granite Mountain, Lucille, Gray Mountain, Hilltop, Maggie Mitchell, Silver Islet, Revenue, Roosevelt, Carbonate King and others. In addition to hardrock mining, are also included are details on gold placer mining in this portion of Colorado. **8.5" X 11", 140 ppgs, Retail Price: $12.99**

Ore Deposits of the London Fault of Colorado - First published in 1941, it has been unavailable since those days and sheds important light on the mines and mineral deposits of the London Fault in Central Colorado's Alma Mining District. This publication sheds important light on the gold veins and lead-silver deposits of the Alma Mining District. Included are geologic details on the London Mine, American Mine, Havigorst Tunnel, Ophir Mine, Mosher Tunnel, London-Butte Mine, Venture Shaft, Hard-To-Beat Mine, Oliver Twist Tunnel, Sacramento Mine, Mudsill Mine, Sherwood Mine, Wagner, Barcoe Tunnel and other mines in this important mining region. 110 ppgs., 10.99

The Mines of Colorado - First published in 1867, it has been unavailable since those days and sheds important light on Colorado's early mining history. Written shortly after the events took place, this publication sheds important light on the Pike's Peak Gold Rush, the discovery of gold on Ralston Creek and Dry Creek in the 1850's, as well as details on the first wave of miners into Colorado and their trials and tribulations as they crossed the Great Plains. Also included are details on early discoveries of lode gold in the mountainous regions of Colorado, details on the early mines hardrock and placer mines, and much more. It is a veritable treasure trove on Colorado's early mining history and will be of great importance to anyone who is interested in the mining of gold or other minerals in Colorado, as well as those interested in the history of the state. 478 ppgs., 29.99

The La Plata Mining District of Colorado - Originally titled "Geology and Ore Deposits in the Vicinity of the La Plata District of Colorado" and first published in 1949, it has been unavailable since those days and sheds important light on the mines and mineral deposits of the La Plata Mining District of Colorado. 214 ppgs., 19.99

Washington Mining Books

The Republic Mining District of Washington - Unavailable since 1910, this important publication was originally published by the Washington Geologic Survey and has been unavailable for a century. Topics include the geology, rock formations and the formation of ore deposits in this important mining area of Washington State. Also included are hard to find details on the geology, history and locations of dozens of mines in the area. Some of the mines featured include the New Republic Mine, Ben Hur, Morning Glory, the South Republic Mine, Quilp, Surprise, Black Tail, Lone Pine, San Poil, Mountain Lion, Tom Thumb, Elcaliph and many others. **8.5" X 11", 94 ppgs, Retail Price: $10.99**

The Myers Creek and Nighthawk Mining Districts of Washington - Unavailable since 1911, this important publication was originally published by the Washington Geologic Survey and has been unavailable for a century. Topics include the geology, rock formations and the formation of ore deposits in these important mining areas of Washington State. Also included are hard to find details on the geology, history and locations of dozens of mines in the area. Some of the mines featured include the Grant Mine, Monterey, Nip and Tuck, Myers Creek, Number Nine, Neutral, Rainbow, Aztec, Crystal Butte, Apex, Butcher Boy, Molson, Mad River, Olentangy, Delate, Kelsey, Golden Chariot, Okanogan, Ohio, Forty-Ninth Parallel, Nighthawk, Favorite, Little Chopaka, Summit, Number One, California, Peerless, Caaba, Prize Group, Ruby, Mountain Sheep, Golden Zone, Rich Bar, Similkameen, Kimberly, Triune, Hiawatha, Trinity, Hornsilver, Maquae, Bellevue, Bullfrog, Palmer Lake, Ivanhoe, Copper World and many others. **8.5" X 11", 136 ppgs, Retail Price: $12.99**

The Blewett Mining District of Washington - Unavailable since 1911, this important publication was originally published by the Washington Geologic Survey and has been unavailable for a century. Topics include the geology, rock formations and the formation of ore deposits in this important mining area of Washington State. Also included are hard to find details on the geology, history and locations of dozens of mines in the area. Some of the mines featured include the Washington Meteor, Alta Vista, Pole Pick, Blinn, North Star, Golden Eagle, Tip Top, Wilder, Golden Guinea, Lucky Queen, Blue Bell, Prospect, Homestake, Lone Rock, Johnson, and others. **8.5" X 11", 134 ppgs, Retail Price: $12.99**

Silver Mining In Washington - Unavailable since 1955, this important publication was originally published by the Washington Geologic Survey. Featured are the hard to find locations and details pertaining to Washington's silver mines. **8.5" X 11", 180 ppgs, Retail Price: $15.99**

The Mines of Snohomish County Washington - Unavailable since 1942, this important publication was originally published by the Washington Geologic Survey and has been unavailable for seventy years. Featured are details on a large number of gold, silver, copper, lead and other metallic mineral mines. Included are the locations of each historic mine, along with information on the commodity produced. **8.5" X 11", 98 ppgs, Retail Price: $10.99**

The Mines of Chelan County Washington - Unavailable since 1943, this important publication was originally published by the Washington Geologic Survey and has been unavailable for seventy years. Featured are details on a large number of gold, silver, copper, lead and other metallic mineral mines. Included are the locations of each historic mine, along with information on the commodity. **8.5" X 11", 88 ppgs, Retail Price: $9.99**

Metal Mines of Washington - Unavailable since 1921, this important publication was originally published by the Washington Geologic Survey and has been unavailable for nearly ninety years. Widely considered a masterpiece on the Washington Mining Industry, "Metal Mines of Washington" sheds light on the important details of Washington's early mining years. Featured are details on hundreds of gold, silver, copper, lead and other metallic mineral mines. Included are hard to find details on the mineral resources of this state, as well as the locations of historic mines. Lavishly illustrated with maps and historic photos and complete with a glossary to explain any technical terms found in the text, this is one of the most important works on mining in the State of Washington. No prospector or miner should be without it if they are interested in mining in Washington. **8.5" X 11", 396 ppgs, Retail Price: $24.99**

Gem Stones In Washington - Unavailable since 1949, this important publication was originally published by the Washington Geologic Survey and has been unavailable since first published. Included are details on where to find naturally occurring gem stones in the State of Washington, including quartz crystal, amethyst, smoky quartz, milky quartz, agates, bloodstone, carnelian, chert, flint, jasper, onyx, petrified wood, opal, fire opal, hyalite and others. **8.5" X 11", 54 ppgs, Retail Price: $8.99**

The Covada Mining District of Washington - Unavailable since 1913, this important publication was originally published by the Washington Geologic Survey and has been unavailable for a century. Topics include the geology, rock formations and the formation of ore deposits in this important mining area of Washington State. Also included are hard to find details on the geology, history and locations of dozens of mines in the area. Some of the mines featured include the Admiral, Advance, Algonkian, Big Bug, Big Chief, Big Joker, Black Hawk, Black Tail, Black Thorn, Captain, Cherokee Strip, Colorado, Dan Patch, Dead Shot, Etta, Good Ore, Greasy Run, Great Scott, Idora, IXL, Jay Bird, Kentucky Bell, King Solomon, Laurel, Laura S, Little Jay, Meteor, Neglected, Northern Light, Old Nell, Plymouth Rock, Polaris, Quandary, Reserve, Shoo Fly, Silver Plume, Three Pines, Vernie, White Rose and dozens of others. **8.5" X 11", 114 ppgs, Retail Price: $10.99**

The Index Mining District of Washington - Unavailable since 1912, this important publication was originally published by the Washington Geologic Survey and has been unavailable for a century. Topics include the geology, rock formations and the formation of ore deposits in this important mining area of Washington State. Also included are hard to find details on the geology, history and locations of dozens of mines in the area. Some of the mines featured include the Sunset, Non-Pareil, Ethel Consolidated, Kittaning, Merchant, Homestead, Co-operative, Lost Creek, Uncle Sam, Calumet, Florence-Rae, Bitter Creek, Index Peacock, Gunn Peak, Helena, North Star, Buckeye. Copper Bell, Red Cross and others. **8.5" X 11", 114 ppgs, Retail Price: $11.99**

Mining & Mineral Resources of Stevens County Washington - Unavailable since 1920, this important publication was originally published by the Washington Geologic Survey and has been unavailable for a century. Topics include the geology, rock formations and the formation of ore deposits in these important mining areas of Washington State. Also included are hard to find details on the geology, history and locations of hundreds of mines in the area. **8.5" X 11", 372 ppgs, Retail Price: $24.99**

The Mines and Geology of the Loomis Quadrangle Okanogan County, Washington - Unavailable since 1972, this important publication was originally published by the Washington Geologic Survey and has been unavailable for a century. Topics include the geology, rock formations and the formation of ore deposits in this important mining area of Washington State. Also included are hard to find details on the geology, history and locations of dozens of gold, copper, silver and other mines in the area. **8.5" X 11", 150 ppgs, Retail Price: $12.99**

The Conconully Mining District of Okanogan County Washington - Unavailable since 1973, this important publication was originally published by the Washington Geologic Survey and has been unavailable for a century. Topics include the geology, rock formations and the formation of ore deposits in this important mining area of Washington State, which also includes Salmon Creek, Blue Lake and Galena. Also included are hard to find details on the geology, mining history and locations of dozens of mines in the area. Some of the mines include Arlington, Fourth of July, Sonny Boy, First Thought, Last Chance, War Eagle-Peacock, Wheeler, Mohawk, Lone Star, Woo Loo Moo Loo, Keystone, Hughes, Plant-Callahan, Johnny Boy, Leuena, Gubser, John Arthur, Tough Nut, Homestake, Key and many others **8.5" X 11", 68 ppgs, Retail Price: $8.99**

Wyoming Mining Books

Mining in the Laramie Basin of Wyoming - Unavailable since 1909, this publication was originally compiled by the United States Department of Interior. Also included are insights into the mineralization and other characteristics of this important mining region, especially in regards to coal, limestone, gypsum, bentonite clay, cement, sand, clay and copper. **8.5" X 11", 104 ppgs, Retail Price: $11.99**

New Mexico Mining Books

The Mogollon Mining District of New Mexico - Unavailable since 1927, this important publication was originally published by the US Department of Interior and has been unavailable for 80 years. Topics include the geology, rock formations and the formation of ore deposits in this important mining area in New Mexico. Of particular focus is information on the history and production of the ore deposits in this area, their form and structure, vein filling, their paragenesis, origins and ore shoots, as well as oxidation and supergene enrichment. Also included are hard to find details, including the descriptions and locations of numerous gold, silver and other types of mines, including the Eureka, Pacific, South Alpine, Great Western, Enterprise, Buffalo, Mountain View, Floride, Gold Dust, Last Chance, Deadwood, Confidence, Maud S., Deep Down, Little Fanney, Trilby, Johnson, Alberta, Comet, Golden Eagle, Cooney, Queen, the Iron Crown, Eberle, Clifton, Andrew Jackson mine, Mascot and others. **8.5" X 11", 144 ppgs, Retail Price: $12.99**

The Percha Mining District of Kingston New Mexico - Unavailable since 1883, this important publication was originally published by the Kingston Tribune and has been unavailable for over one hundred and thirty five years. Having been written during the earliest years of gold and silver mining in the Percha Mining District, unlike other books on the subject, this work offers the unique perspective of having actually been written while the early mining history of this area was still being made. In fact, the work was written so early in the development of this area that many of the notable mines in the Percha District were less than a few years old and were still being operated by their original discoverers with the same enthusiasm as when they were first located. Included are hard to find details on the very earliest gold and silver mines of this important mining district near Kingston in Sierra County, New Mexico. **8.5" X 11", 68 ppgs, Retail Price: $9.99**

East Coast Mining Books

<u>The Gold Fields of the Southern Appalachians</u> - Unavailable since 1895, this important publication was originally published by the US Department of Interior and has been unavailable for nearly 120 years. Topics include the geology, rock formations and the formation of ore deposits in this important mining area of the American South. Of particular focus is information on the history and statistics of the ore deposits in this area, their form and structure and veins. Also included are details on the placer gold deposits of the region. The gold fields of the Georgian Belt, Carolinian Belt and the South Mountain Mining District of North Carolina are all treated in descriptive detail. Included are hard to find details, including the descriptions and locations of numerous gold mines in Georgia, North Carolina and elsewhere in the American South. Also included are details on the gold belts of the British Maritime Provinces and the Green Mountains. **8.5" X 11", 104 ppgs, Retail Price: $9.99**

Gold Rush Tales Series

Millions in Siskiyou County Gold - In this first volume of the "Gold Rush Tales" series, leading mining historian and editor Kerby Jackson, introduces us to the story of how millions of dollars worth of gold was discovered in Siskiyou County during the California Gold Rush. Lavishly illustrated with photos from the 19th Century, this hard to find information was first published in 1897 and sheds important light onto the gold rush era in Siskiyou County, California and the experiences of the men who dug for the gold and actually found it. **8.5" X 11", 82 ppgs, Retail Price: $9.99**

The California Rand in the Days of '49 - In this second volume of the "Gold Rush Tales" series, leading mining historian and editor Kerby Jackson, introduces us to four tales from the California Gold Rush. Lavishly illustrated with photos from the 19th Century, this hard to find information was first published in 1890's and includes the stories of "California's Rand", details about Chinese miners, how one early miner named Baker struck it rich and also the story of Alphonzo Bowers, who invented the first hydraulic gold dredge. **8.5" X 11", 54 ppgs, Retail Price: $9.99**

More Mining Books

Prospecting and Developing A Small Mine - Topics covered include the classification of varying ores, how to take a proper ore sample, the proper reduction of ore samples, alluvial sampling, how to understand geology as it is applied to prospecting and mining, prospecting procedures, methods of ore treatment, the application of drilling and blasting in a small mine and other topics that the small scale miner will find of benefit. **8.5" X 11", 112 ppgs, Retail Price: $11.99**

Timbering For Small Underground Mines - Topics covered include the selection of caps and posts, the treatment of mine timbers, how to install mine timbers, repairing damaged timbers, use of drift supports, headboards, squeeze sets, ore chute construction, mine cribbing, square set timbering methods, the use of steel and concrete sets and other topics that the small underground miner will find of benefit. This volume also includes twenty eight illustrations depicting the proper construction of mine timbering and support systems that greatly enhance the practical usability of the information contained in this small book. **8.5" X 11", 88 ppgs. Retail Price: $10.99**

Timbering and Mining - A classic mining publication on Hard Rock Mining by W.H. Storms. Unavailable since 1909, this rare publication provides an in depth look at American methods of underground mine timbering and mining methods. Topics include the selection and preservation of mine timbers, drifting and drift sets, driving in running ground, structural steel in mine workings, timbering drifts in gravel mines, timbering methods for driving shafts, positioning drill holes in shafts, timbering stations at shafts, drainage, mining large ore bodies by means of open cuts or by the "Glory Hole" system, stoping out ore in flat or low lying veins, use of the "Caving System", stoping in swelling ground, how to stope out large ore bodies, Square Set timbering on the Comstock and its modifications by California miners, the construction of ore chutes, stoping ore bodies by use of the "Block System", how to work dangerous ground, information on the "Delprat System" of stoping without mine timbers, construction and use of headframes and much more. This volume provides a reference into not only practical methods of mining and timbering that may be employed in narrow vein mining by small miners today, but also rare insights into how mines were being worked at the turn of the 19th Century. **8.5" X 11", 288 ppgs. Retail Price: $24.99**

A Study of Ore Deposits For The Practical Miner - Mining historian Kerby Jackson introduces us to a classic mining publication on ore deposits by J.P. Wallace. First published in 1908, it has been unavailable for over a century. Included are important insights into the properties of minerals and their identification, on the occurrence and origin of gold, on gold alloys, insights into gold bearing sulfides such as pyrites and arsenopyrites, on gold bearing vanadium, gold and silver tellurides, lead and mercury tellurides, on silver ores, platinum and iridium, mercury ores, copper ores, lead ores, zinc ores, iron ores, chromium ores, manganese ores, nickel ores, tin ores, tungsten ores and others. Also included are facts regarding rock forming minerals, their composition and occurrences, on igneous, sedimentary, metamorphic and intrusive rocks, as well as how they are geologically disturbed by dikes, flows and faults, as well as the effects of these geologic actions and why they are important to the miner. Written specifically with the common miner and prospector in mind, the book will help to unlock the earth's hidden wealth for you and is written in a simple and concise language that anyone can understand. **8.5" X 11", 366 ppgs. Retail Price: $24.99**

Mine Drainage - Unavailable since 1896, this rare publication provides an in depth look at American methods of underground mine drainage and mining pump systems. This volume provides a reference into not only practical methods of mining drainage that may be employed in narrow vein mining by small miners today, but also rare insights into how mines were being worked at the turn of the 19th Century. **8.5" X 11", 218 ppgs. Retail Price: $24.99**

Fire Assaying Gold, Silver and Lead Ores - Unavailable since 1907, this important publication was originally published by the Mining and Scientific Press and was designed to introduce miners and prospectors of gold, silver and lead to the art of fire assaying. Topics include the fire assaying of ores and products containing gold, silver and lead; the sampling and preparation of ore for an assay; care of the assay office, assay furnaces; crucibles and scorifiers; assay balances; metallic ores; scorification assays; cupelling; parting' crucible assays, the roasting of ores and more. This classic provides a time honored method of assaying put forward in a clear, concise and easy to understand language that will make it a benefit to even beginners. **8.5" X 11", 96 ppgs. Retail Price: $11.99**

Methods of Mine Timbering - Originally published in 1896, this important publication on mining engineering has not been available for nearly a century. Included are rare insights into historical methods of timbering structural support that were used in underground metal mines during the California that still have a practical application for the small scale hardrock miner of today. **8.5" X 11", 94 ppgs. Retail Price: $10.99**

The Enrichment of Copper Sulfide Ores - First published in 1913, it has been unavailable for over a century. Topics include the definition and types of ore enrichment, the oxidation of copper ores, the precipitation of metallic sulfides. Also included are the results of dozens of lab experiments pertaining to the enrichment of sulfide ores that will be of interest to the practical hard rock mine operator in his efforts to release the metallic bounty from his mine's ore. **8.5" X 11", 92 ppgs. Retail Price: $9.99**

A Study of Magmatic Sulfide Ores - Unavailable since 1914, this rare publication provides an in depth look at magmatic sulfide ores. Some of the topics included are the definition and classification of magmatic ores, descriptions of some magmatic sulfide ore deposits known at the time of publication including copper and nickel bearing pyrrohitic ore bodies, chalcopyrite-bornite deposits, pyritic deposits, magnetite-ileminite deposits, chromite deposits and magmatic iron ore deposits. Also included are details on how to recognize these types of ore deposits while prospecting for valuable hardrock minerals. **8.5" X 11", 138 ppgs. Retail Price: $11.99**

The Cyanide Process of Gold Recovery - Unavailable since 1894 and released under the name "The Cyanide Process: Its Practical Application and Economical Results", this rare publication provides an in depth look at the early use of cyanide leaching for gold recovery from hardrock mine ores. This volume provides a reference into the early development and use of cyanide leaching to recover gold. **8.5" X 11", 162 ppgs. Retail Price: $14.99**

California Gold Milling Practices - Unavailable since 1895 and released under the name "California Gold Practices", this rare publication provides an in depth look at early methods of milling used to reduce gold ores in California during the late 19th century. This volume provides a reference into the early development and use of milling equipment during the earliest years of the California Gold Rush up to the age of the Industrial Revolution. Much of the information still applies today and will be of use to small scale miners engaging in hardrock mining. **8.5" X 11", 104 ppgs. Retail Price: $10.99**

Leaching Gold and Silver Ores With The Plattner and Kiss Processes - Mining historian Kerby Jackson introduces us to a classic mining publication on the evaluation and examination of mines and prospects by C.H. Aaron. First published in 1881, it has been unavailable for over a century and sheds important light on the leaching of gold and silver ores with the Plattner and Kiss processes. **8.5" X 11", 204 ppgs. Retail Price: $15.99**

The Metallurgy of Lead and the Desilverization of Base Bullion - First published in 1896, it has been unavailable for over a century and sheds important light on the the recovery of silver from lead based ores. Some of the topics include the properties of lead and some of its compounds, lead ores such as galenite, anglesite, cerussite and others, the distribution of lead ores throughout the United States and the sampling and assaying of lead ores. Also covered is the metallurgical treatment of lead ores, as well as the desilverization of lead by the Pattinson Process and the Parkes Process. Hofman's text has long been considered one of the most important early works on the recovery of silver from lead based ores. 8.5" X 11", 452 ppgs. **Retail Price: $29.99**

Ore Sampling For Small Scale Miners - First published in 1916, it has been unavailable for over a century and sheds important light on historic methods of ore sampling in hardrock mines. Topics include how to take correct ore samples and the conditions that affect sampling, such as their subdivision and uniformity. Particular detail is given to methods of hand sampling ore bodies by grab sample, pipe sample and coning, as well as sampling by mechanical methods. Also given are insights into the screening, drying and grinding processes to achieve the most consistent sample results and much more. 8.5" X 11", 124 ppgs. **Retail Price: $12.99**

The Extraction of Silver, Copper and Tin from Ores - First published in 1896, it has been unavailable for over a century and sheds important light on how historic miners recovered silver, copper and tin from their mining operations. The book is split into three sections, including a discussion on the Lixiviation of Silver Ores, the mining and treatment of copper ores as practiced at Tharsis, Spain and the smelting of tin as it was practiced by metallurgists at Pulo Brani, Singapore. Also included is an overview and analysis of these historic metal recovery methods that will be of benefit to those interested in the extraction of silver, copper and tin from small mines. 8.5" X 11", 118 ppgs. **Retail Price: $14.99**

The Roasting of Gold and Silver Ores - First published in 1880, it has been unavailable for over a century and sheds important light on how historic miners recovered gold and silver rom their mining operations. Topics include details on the most important silver and free milling gold ores, methods of desulphurization of ores, methods of deoxidation, the chlorination of ores, methods and details on roasting gold and silver ores, notes on furnaces and more. Also included are details on numerous methods of gold and silver recovery, including the Ottokar Hofman's Process, the Patera Process, Kiss Process, Augustin Process, Ziervogel Process and others. 8.5" X 11", 178 ppgs. **Retail Price: $19.99**

The Examination of Mines and Prospects - First published in 1912, it has been unavailable for over a century and sheds important light on how to examine and evaluate hardrock mines, prospects and lode mining claims. Sections include Mining Examinations, Structural Geology, Structural Features of Ore Deposits, Primary Ores and their Distribution, Types of Primary Ore Deposits, Primary Ore Shoots, The Primary Alteration of Wall Rocks, Alterations by Surface Agencies, Residual Ores and their Distribution, Secondary Ores and Ore Shoots and Vein Outcrops. This hard to find information is a must for those who are interested in owning a mine or who already own a lode mining claim and wish to succeed at quartz mining. 8.5" X 11", 250 ppgs. **Retail Price: $19.99**

Garnets: Their Mining, Milling and Utilization - First published in 1925, it has been unavailable since those days and sheds important light on the mining, milling and utilization of garnets. Included are details on the characteristics of garnets, where they are found and how they were mined. 78 ppgs, 10.99

Gemstones and Precious Stones of North America - Leading mining historian Kerby Jackson introduces us to a classic mining publication on the gems and precious stones of the United States, Canada and mexico. First published in 1890, it has been unavailable since those days and sheds important light on the gems and precious stones that may be found in North America. Included are chapters on diamonds, corundum, sapphire, ruby, topaz, emerald, disapore, spinel, turquoise, tourmaline, garnets, beyrl, peridot, zircon, quartz crystals, feldspars, pearls and many others. Included are details on where these gems and precious stones may be found throughout North America, as well as their characteristics. 360 ppgs, 24.99

Mining Camps and Mining Districts - First released in 1885 by Charles Howard Shinn under the title "Mining Camps: A Study in American Frontier Government", this publication offers a unique look at how early gold miners established their own forms of representative government during the California Gold Rush. Drawing on the the early mining codes of mideviel German miners in the Harz Mountains, on the mining customs of the Cornish tin miners and early Spanish mining laws introduced into California, the miners established the first governments in the American West. 340 ppgs, 24.99

BLM Field Handbook for Mineral Examiners - Leading mining historian Kerby Jackson introduces us to a classic mining publication on mine evaluation. First published in 1962, this work sheds important light on the techniques of BLM Mineral Examiners to perform validity on mining claims. 132 ppgs, 10.99

Six Months In The Gold Mines During The California Gold Rush - Unavailable since 1850, this important work is a first hand account of one "49'ers" personal experience during the great California Gold Rush, shedding important light on one of the most exciting periods in the history of not only California, but also the world. Compiled from journals written between 1847 and 1849 by E. Gould Buffum, a native of New York, "Six Months In The Gold Mines During The California Gold Rush" offers a rare look into the day to day lives of the people who came to California to work in her gold mines when the state was still a great frontier. **8.5" X 11", 290 ppgs. Retail Price: $19.99**

The Discovery of Gold in Australia - First published in 1852, it has been unavailable since those days and sheds important light on Australia's gold mining history. Included are rare communications between British agents and the British Crown when gold was first discovered in Australia in 1851. This rare text contains hard to find details on Australia's first mining camps and Britain's early attempts to provide for the orderly regulation of gold mines in that part of the world. Also of interest are hard to find extracts of articles that appeared in the early colonial newspapers that did their best to report on Australia's gold rush as it took place.
102 ppgs, 10.99